Springer Complexity

Springer Complexity is an interdisciplinary program publishing the best research and academic-level teaching on both fundamental and applied aspects of complex systems – cutting across all traditional disciplines of the natural and life sciences, engineering, economics, medicine, neuroscience, social and computer science.

Complex Systems are systems that comprise many interacting parts with the ability to generate a new quality of macroscopic collective behavior the manifestations of which are the spontaneous formation of distinctive temporal, spatial or functional structures. Models of such systems can be successfully mapped onto quite diverse "real-life" situations like the climate, the coherent emission of light from lasers, chemical reaction-diffusion systems, biological cellular networks, the dynamics of stock markets and of the internet, earthquake statistics and prediction, freeway traffic, the human brain, or the formation of opinions in social systems, to name just some of the popular applications.

Although their scope and methodologies overlap somewhat, one can distinguish the following main concepts and tools: self-organization, nonlinear dynamics, synergetics, turbulence, dynamical systems, catastrophes, instabilities, stochastic processes, chaos, graphs and networks, cellular automata, adaptive systems, genetic algorithms and computational intelligence.

The three major book publication platforms of the Springer Complexity program are the monograph series "Understanding Complex Systems" focusing on the various applications of complexity, the "Springer Series in Synergetics", which is devoted to the quantitative theoretical and methodological foundations, and the "SpringerBriefs in Complexity" which are concise and topical working reports, case-studies, surveys, essays and lecture notes of relevance to the field. In addition to the books in these two core series, the program also incorporates individual titles ranging from textbooks to major reference works.

Editorial and Programme Advisory Board

Understanding Complex Systems

Founding Editor: S. Kelso

Future scientific and technological developments in many fields will necessarily depend upon coming to grips with complex systems. Such systems are complex in both their composition – typically many different kinds of components interacting simultaneously and nonlinearly with each other and their environments on multiple levels – and in the rich diversity of behavior of which they are capable.

The Springer Series in Understanding Complex Systems series (UCS) promotes new strategies and paradigms for understanding and realizing applications of complex systems research in a wide variety of fields and endeavors. UCS is explicitly transdisciplinary. It has three main goals: First, to elaborate the concepts, methods and tools of complex systems at all levels of description and in all scientific fields, especially newly emerging areas within the life, social, behavioral, economic, neuro- and cognitive sciences (and derivatives thereof); second, to encourage novel applications of these ideas in various fields of engineering and computation such as robotics, nano-technology and informatics; third, to provide a single forum within which commonalities and differences in the workings of complex systems may be discerned, hence leading to deeper insight and understanding.

UCS will publish monographs, lecture notes and selected edited contributions aimed at communicating new findings to a large multidisciplinary audience.

More information about this series at http://www.springer.com/series/5394

Dean S. Hartley III

Unconventional Conflict

A Modeling Perspective

Springer

Dean S. Hartley III
Hartley Consulting
Oak Ridge, Tennessee
USA

ISSN 1860-0832 ISSN 1860-0840 (electronic)
Understanding Complex Systems
ISBN 978-3-319-51934-0 ISBN 978-3-319-51935-7 (eBook)
DOI 10.1007/978-3-319-51935-7

Library of Congress Control Number: 2017930967

Printed on acid-free paper

This Springer imprint is published by Springer Nature
The registered company is Springer International Publishing AG
The registered company address is: Gewerbestrasse 11, 6330 Cham, Switzerland

Foreword

This book represents a certain amount of work in its compilation; however, for me, it was more of a pleasure than a chore. Those I asked to review it and make comments are the ones who did the real work.

The first group of people and organizations deserving thanks are the ones who owned the problems and supported the work that I did on those problems.

- Dave Haut, Lynda Jaques, and the U.S. Pacific Command (USPACOM) introduced me to the problems of unconventional conflict and supported my earliest work.
- Jackie Henningsen and the Office of the Secretary of Defense (OSD), Program Analysis and Evaluation (PA&E), supported additional work that helped define the problems in unconventional conflict.
- Kevin Baugh and Moses Kamai at the Office of the Assistant Secretary of Defense (OASD), Special Operations/Low Intensity Conflict (SO/LIC), also supported additional work that helped define the problems in unconventional conflict.
- Forrest Crain, Phil Barry, Mike Rugienius, and the Defense Modeling and Simulation Office (DMSO) supported work in creating systems of models of unconventional conflict and the beginning of the definition of verification, validation, and accreditation (VV&A) for such models.
- Jon Lockey and The Joint Staff/J8 provided a real-world study in which to use the system of models.
- Alexander Kott and the Defense Advanced Research Projects Agency (DARPA) provided the opportunity to codify VV&A of unconventional conflict models.
- Richard Pei, Gary Citrenbaum, and the U.S. Army Communications-Electronics Research Development and Engineering Center (CERDEC) Intelligence and Information Warfare Directorate (I2WD) provided the opportunity to apply the VV&A process.

- Mike Baranick and the Center for Technology and National Security Policy at the National Defense University (NDU) hosted a series of workshops relating to unconventional conflict.
- Loren Cobb at the University of Colorado also provided the opportunity to apply the VV&A process.
- Paul Works and the U.S. Army's Training and Doctrine Command (TRADOC) Analysis Center (TRAC) provided the opportunity to develop an ontology that applies to unconventional conflict.

The second group of people deserving thanks is comprised of the coworkers on these projects. These included the following:

Myriam Abramson	Lee Hively	B. Rigdon
Gary Austin	Dave Holdsworth	Frederick Schaaf
Rich Bell	Mike Hopkins	Al Sciarretta
Ed Bitinas	Kara Kruse	Bob Senko
Curtis Blais	Lee Lacy	David Signori
R. Breuer	Andy Loebl	Mike Sovereign
John Cipparone	Mike Loesekann	Stuart Starr
Doug Clark	Shu Ping Lu	Mike Strycharz
Robert Clemence	John Martellaro	Ben Thomas
Mark Crooks	Jimmie McEver	David Thompson
John Denson	John Mericle	Victoria Turley
Christian Farrell	Donna Middleton	B. P. van Leeuwen
Mike Farrell	Rick Montera	Howard Whitley
Christian Gauvin	David Noble	Laura Whitney
R. Harrigan	Steve Packard	Jordan Wilcox
Bruce Harris	Patricia Partnow	Alexander Woodcock
Zoe Henscheid	John Quillinan	
Roger Hillson	Wayne Randolph	

Finally, I thank those who reviewed this book and made comments: Lee Lacy, Steve Packard, and, most especially, Mike McCurdy. Errors and omissions are my responsibility, as I do not always accept the advice I am given.

Preface

The simplest thing that can be said about unconventional conflict is that it is not conventional war. It may include combat operations. It may include multiple conflicting parties. It may have a time span measured in years, depending on the situation. It may be confined to a single country or span a continent. It certainly includes social and cultural behavior issues, and the parties to the conflict may not all play by the same set of rules.

This volume describes issues in modeling unconventional conflict and suggests a new way to do the modeling. With the creation of a holistic ontology that describes the domain, we now have a nearly complete specification for modeling unconventional conflict at the theater level. Further, this new approach disentangles the more straightforward modeling issues from the harder problems of discovering and selecting the social theories that are required to justify the connections between causes and effects.

Even with the ontological approach to modeling unconventional conflict, the modeling process will be complex. And, even with the disentanglement of the "hard parts" from the easier parts, there is no guarantee that the hard parts will be modeled correctly.

However, it is important to realize that as long as unconventional conflicts take place, someone has to deal with them. Without the support of a coded model, that someone must rely on his or her mental model of the situation and the likely responses of other parties in making his or her decisions. Just as Napoleon was regarded as a strategic genius in executing the conventional war of the time, there may be a genius in charge of conducting operations in an unconventional conflict. But if there are geniuses, there are many more "average" strategists and some awful strategists. There is no guarantee that we will have a genius when we need one.

A coded model has the advantage over a mental model of being specified, so that if it is seen to be in error, that error can be identified, addressed, and mitigated, if not fully corrected. Further, in using a mental model, it is easy to forget to consider something. The coded model will, at least, consider everything that is specified—every time.

Chapter 1 introduces the book by presenting some ways of thinking about unconventional conflict and some modeling preliminaries and an expanded discussion of the software design process.

Before discussing how to model unconventional conflict, it helps to have a common view of what unconventional conflict is. Chapter 2 discusses the many aspects of unconventional conflict and some past analysis efforts.

Although modeling unconventional conflict is new when compared to efforts to model combat, there have been a number of models constructed in the past two decades and a number of people have put a lot of thought into the matter. Chapter 3 discusses modeling constructs and approaches and past modeling of unconventional conflict in some detail.

One of the major developments in the thinking about unconventional conflict has been the creation of the DIME/PMESII paradigm. The central concepts are that the state of a situation can and should be described and the descriptive variables can be organized into Political, Military, Economic, Social, Information, and Infrastructure (PMESII) categories (or some variant). Similarly, the efforts of a nation to change the state of the situation can be organized into Diplomatic, Informational, Military, and Economic (DIME) levers of national power (or some variant). Chapter 4 concentrates on the DIME/PMESII paradigm.

A second major development has been the creation of the concept of ontologies as a means of describing what we know about a domain and the application of this concept to unconventional conflict. Chapter 5 introduces ontologies and discusses the application of this concept to unconventional conflict.

The third major development started with combat modeling—the formalization of the concepts of verification, validation, and accreditation (VV&A). It culminated with the application of VV&A to unconventional conflict models. Chapter 6 discusses verification, validation, and accreditation in general and its particular application to models of unconventional conflict.

With the preliminaries in place, Chapter 7 introduces a prescription for modeling unconventional conflict (within the geographical extent, level of resolution, and duration under consideration here).

Chapter 8 reviews the contents of the previous chapters and recaps the major points that have been made.

The front matter includes a listing of the acronyms used in this book and their definitions. The end matter includes a bibliography of the citations in the text and an index to the appearance of important terms in the text.

I have written this book from a U.S. perspective. Despite having performed some work for the North Atlantic Treaty Organization (NATO), my personal perspective is basically a U.S. perspective. Many other countries are involved in unconventional conflicts and may find this work useful because of the wide variety of unconventional conflicts in which the U.S.A. has been involved.

Significant parts of this book arose from work performed for the Defense Modeling and Simulation Office (DMSO) of the U.S. Department of Defense (DoD) and other U.S. government organizations. DMSO is now known as the Modeling & Simulation Coordination Office (M&SCO); however, its vision is the same, to support the full spectrum of DoD's activities and operations with improved modeling and simulation (M&S) capabilities. I hope that this book will serve to improve M&S capabilities, also; however, no part of the U.S. government was involved in writing this book and the views expressed here are solely my own.

Contents

Acronyms

AAR	Actor-action-result
ABM	Agent-based model
ACT	Analysis of Complex Threats
ADS	Authoritative data source
AMORS	Asia-Pacific Military Operations Research Symposium
AoR	Area of responsibility
APSO	Aggravated Peace Support Operations
CASOS	Computational Analysis of Social and Organizational Systems
CATS	Consequence Assessment Toolset
CD	Counter-drug
CERDEC	Communications-Electronics Research Development and Engineering Center
CFLD	Canadian Forces Landmine Database
CI	Counterinsurgency
CIA	Central Intelligence Agency
CIFP	Country indicators for foreign policy
CJCS	Chairman of the Joint Chiefs of Staff
CM	Conceptual model
CMOC	Civil Military Operations Center
CMPO	Conceptual Model of Peace Operations
COA	Course of action
COG	Center of gravity
COI	Community of interest
COIN	Counterinsurgency
COMPOEX	Conflict Modeling, Planning and Outcomes Experimentation
CONOPS	Concept of operations
CONT	Military contingency
CT	Counterterrorism
DARPA	Defense Advanced Research Projects Agency
DET	Data Extraction Tool

DEXES	Deployable Exercise Support System
DEXES/CAM	DEXES Civil Affairs Module
DIA	Defense Intelligence Agency
DIAMOND	Diplomatic and Military Operations in a Non-warfighting Domain
DIAMOND-US	DIAMOND, U.S. version
DIF	Data interchange format
DIME	Diplomatic, Information, Military, Economic
DIMEFIL	DIME plus Financial, Intelligence, Law Enforcement
DMSO	Defense Modeling and Simulation Office
DoD	Department of Defense
DOJ	Department of Justice
DOS	Department of State
DP	Dimensional parameter
DR	Disaster Relief
DRC	Dynamics Research Corporation
DSTL	Defence Science and Technology Laboratory
EBR	Evidence Based Research
EP-POL	Electric power-petroleum oil lubricant
ESS	Enhanced Strategic Success
FASP	Foreign Assistance Standardized Program
FAST	Flexible Asymmetric Simulation Technologies
FEMA	Federal Emergency Management Agency
FID	Foreign Internal Defense
FON	Freedom of Navigation
GAO	General Accounting Office, Government Accountability Office
GMU	George Mason University
GTD	Global Terrorism Database
GTO	Goal-task-owner
GUI	Graphical user interface
GWOT	Global War on Terrorism
HA	Humanitarian Assistance
HA/DR	Humanitarian Assistance and Disaster Relief
HITL	Human-in-the-loop
HN	Host nation
HPMF	HSCB PMESII Model Framework
HSCB	Human, social, cultural behavior
HUMINT	Human Intelligence
I2WD	Intelligence and Information Warfare Directorate
ICEWS	Integrated Crisis Early Warning System
ID	Influence diagram
IED	Improvised explosive device
IO	International Organization
ISAF	International Security Assistance Force

ISIS	Islamic State of Iraq and Syria
ISSM	Interim Semi-static Stability Model
ITERATE	International Terrorism: Attributes of Terrorist Events
IW	Irregular Warfare
JCATS	Joint Conflict and Tactical Simulation
JFAST	Joint Flow and Analysis System for Transportation
JFC	Joint force commander
JUORS	Japan-US Operations Research Seminar
JWAC	Joint Warfare Analysis Center
JWARS	Joint Warfare Simulation
LIC	Low-Intensity Conflict
LLIW	Lessons Learned Information Warehouse
M&S	Modeling and Simulation
M&SCO	Modeling & Simulation Coordination Office
MCO	Military Contingency Operations
MIDAS	Model for Inter-theater Deployment by Air and Sea
MIO	Maritime Intercept Operations
MOD	Ministry of Defence
MOE	Measure of Effectiveness
MoFE	Measure of Force Effectiveness
MoM	Measure of Merit
MOOTW	Military Operations Other Than War
MoP	Measure of Performance
MoPE	Measure of Political Effectiveness
MORS	Military Operations Research Society
MPICE	Measuring Progress in Conflict Environments
MRM	Multi-resolution model
MSCA	Military Support to (Domestic) Civil Authorities
MTW	Major Theater War
NATO	North Atlantic Treaty Organization
NATO	SAS NATO System Analysis and Studies
NDU	National Defense University
NEO	Noncombatant Evacuation Operations
NGIC	National Ground Intelligence Center
NGO	Non-Governmental Organization
NI	National Integrity
NPS	Naval Postgraduate School
NSC	National Security Council
OASD	Office of the Assistant Secretary of Defense
OASD (SO/LIC)	OASD Special Operations/Low-Intensity Conflict
OCRS	Office of the Coordinator for Reconstruction and Stabilization
OOTW	Operations Other Than War
OPTEMPO	Operational tempo
OSD	Office of the Secretary of Defense

OSD (PA&E)	OSD Program Analysis and Evaluation
PANDA	Protocol for the Assessment of Nonviolent Direct Action
PC	Personal computer
PE	Peace Enforcement
PK	Peacekeeping
PLU	Price look-up
PMES	Political, Military, Economic, Social
PMESII	Political, Military, Economic, Social, Information, Infrastructure
PMESII-PT	PMESII plus Physical Environment, Time
PMFServ	Performance Moderator Function Server
PO	Peace Operation
PRIME	Probative Rapid Interactive Modeling Environment
PSIR	Predictive Societal Indicators of Radicalism
PSO	Peace Support Operations
PSOM	Peace Support Operations Model
PSYOPS	Psychological Operations
PTSD	Posttraumatic stress disorder
PVO	Private Volunteer Organization
RDECOM	Research, Development and Engineering Command
ROE	Rules of Engagement
ROL	Rule of Law
S&T	Science and Technology
SAR	Search and Rescue
SASO	Stability and Support Operations
SEAS	Synthetic Environments for Analysis and Simulations
SEAS RW	SEAS Reference World
SEED	Simulation Experiments and Efficient Designs
SFP	State Failure Project
SIAM	Situational Influence Assessment Module
SIMVAL	Simulation validation
SKU	Stock keeping unit
SME	Subject matter expert
SOF	Special Operations Forces
SSC	Smaller Scale Contingency
SSTR	Stability, Security, Transition, and Reconstruction
SSTRO	SSTR operations
SWAG	Scientific wild-assed guess
TEO	Task-event-outcome
TRAC	TRADOC Analysis Center
TRADOC	Training and Doctrine Command
TSCP	Theater Security Cooperation Plan
TWG	Tactical War Game
UK	United Kingdom

UN	United Nations
UOB	Unit Order of Battle
UOB DAT	UOB Data Access Tool
U.S.	United States
USAID	U.S. Agency for International Development
USIP	United States Institute of Peace
USPACOM	U.S. Pacific Command
V&V	Verification and validation
VBA	Visual Basic for Applications
VIC	Virtual Information Center
VPMM	Validation Process Maturity Model
VV&A	Verification, validation, and accreditation
WAG	Wild-assed guess
WMD	Weapons of mass destruction
XML	Extensible Markup Language
XMT	XPOD Manipulation Tool
XPOD	XML populated DIF

Chapter 1
Introduction

This book is not about unconventional conflict but about how to model unconventional conflict. In this chapter, we provide an introduction to unconventional conflict and to modeling. Figure 1.1 illustrates the balance between the two topics: unconventional conflict on the left in blue and modeling on the right in orange. Chapters 2 and 3 discuss each subject in more detail. Chapter 4 discusses the DIME/PMESII paradigm, a critical first step in formally describing unconventional conflict. This is followed by a chapter (Chapter 5) on ontologies, which completes the description. Returning to modeling, the next chapter (Chapter 6) discusses verification, validation, and accreditation, which are necessities for modeling anything, but especially important in modeling unconventional conflict. Chapter 7 sketches the process of modeling unconventional conflict that is now possible because of the holistic ontology described in Chapter 5. The final chapter reviews the contents of the previous chapters and recaps the major points that have been discussed.

Span (Geographical Extent)

We are focusing on unconventional conflict at the theater level. Generally, this means that the geographical area under consideration is roughly equivalent to a country. In some cases, two or more countries might comprise the theater, such as Syria and Iraq in the conflict with ISIS. (It is worthwhile to note that from the perspective of Turkey, Turkey plus the Kurdish areas in Syria and Iraq form a different theater, with the conflict being between Turkey and the Kurdish separatists and terrorists (Unal, 2016).)

D.S. Hartley III, *Unconventional Conflict*, Understanding Complex Systems,
DOI 10.1007/978-3-319-51935-7_1

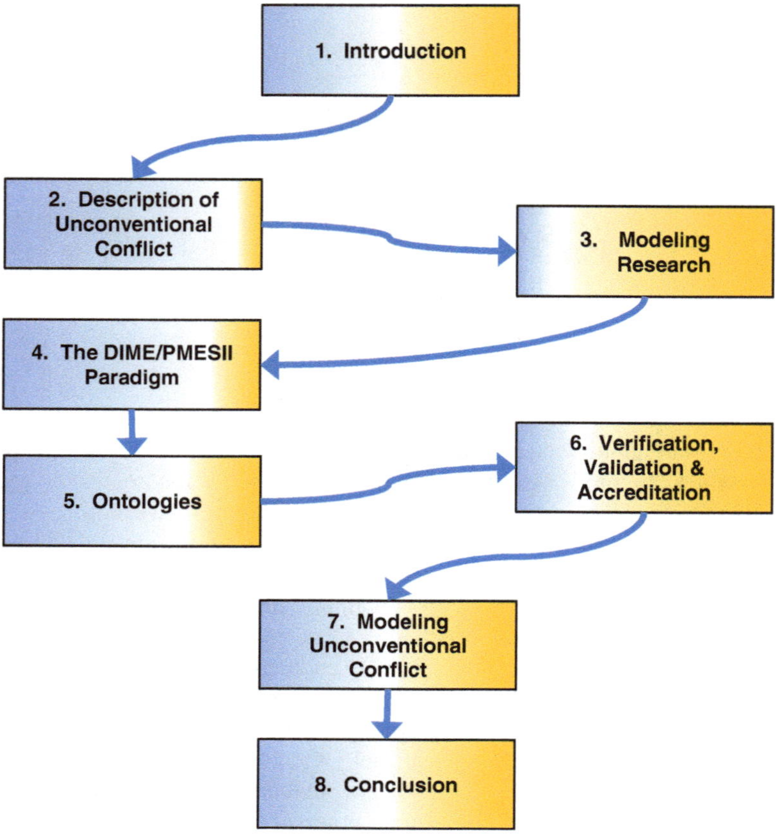

Fig. 1.1 Chapter topics

Level of Resolution

Given a theater perspective, the granularity or level of resolution is restricted by practicality: too fine a granularity requires too many items to enumerate in the model. The number of items in a model affects both the data input and the processing resource requirements. In combat models, the level of resolution is often stated in terms of the normal army unit size (e.g., company, battalion, brigade, or division) that is modeled. The concept is to equate unit size to impact on the model results. Non-army entities are selected for modeling if they have the equivalent impact of the army unit size. In unconventional conflict, army unit size is not a consistently useful measure of impact; however, the concept of defining resolution by impact is still valid. For example, the host nation's legislature (whether in existence or desired to be created) provides a fairly good standard of impact. Another example is provided by non-governmental organizations (NGOs), such as the International Red Cross, that often play major roles in unconventional conflicts. Any other group or person with approximately equivalent impact would have the same level of resolution or granularity.

Certain problems that are relevant to unconventional conflict, such as discerning and dealing with improvised explosive device (IED) networks, require an unsupportable fineness of granularity. Such problems must be dealt with as sub-problems and are not discussed in great detail here.

Time Span (Duration)

Given a theater perspective, the amount of time that will have to be addressed is biased toward denominations in years, rather than days or months. Unconventional conflicts that primarily involve natural disasters tend to be resolved in months; however, unconventional conflicts that primarily involve cultural conflict among multiple sides have often taken years and sometimes decades to resolve.

Defining Unconventional Conflict

We start with the understanding that conflict can range from disagreements between two individuals all the way to global thermonuclear war. One implication of this statement is that the conflict is between or among human beings; however, a broader definition allows for situations in which one of the parties of the conflict is a force of nature, perhaps an erupting volcano, which engenders many of the same actions that would take place in some purely human conflicts.

In this book, we narrow the domain somewhat, omitting nuclear warfare (although not omitting dirty bombs and stolen nuclear weapons) and conflicts that are not of interest to nation states. In this domain, military actions often form a large part of the total activity of the conflict. National militaries may or may not be in charge (often the national diplomats, such as the U.S. Department of State, are in charge of the national operations); however, in many cases, the "heavy lifting" is accomplished by the military, whether in combat roles or in logistic roles. Accordingly, we will use military terminology to describe the operations. There have been many unconventional conflicts in which the United States has not participated; however, this book has been written from a U.S. viewpoint. Thus references to the Departments of Defense and State refer to the U.S. departments and "domestic" refers to U.S. domestic issues. Obviously some conversions would be required to shift the viewpoint to some other country. These conversions should not have any significant impact on the text.

Operations

The horizontal axis of Fig. 1.2 displays several types of military operations.

- The first type of operation, labeled "OOTW," refers to Operations Other than War. This is a complex and diverse set of operations that includes everything a military does other than garrison duty and actual warfare.
- The second type of operation, labeled "SSTR," refers to Stability, Security, Transition, and Reconstruction. SSTR is included in OOTW and involves "various military missions, tasks, and activities conducted outside the United States in coordination with other instruments of national power to maintain or reestablish a safe and secure environment, provide essential governmental services, provide emergency infrastructure reconstruction, and provide humanitarian relief." (Department of Defense, 2009)
- The third type of operation is "Irregular Warfare (IW)." IW "can include any relevant DoD activity and operation such as counterterrorism; unconventional warfare; foreign internal defense; counterinsurgency; and stability operations that, in the context of IW, involve establishing or reestablishing order in a fragile state or territory." (Department of Defense, 2014)
- IW overlaps both OOTW and the fourth type of operation, "Conventional Warfare," which includes warfare as it was known from World War I through the Korean War.

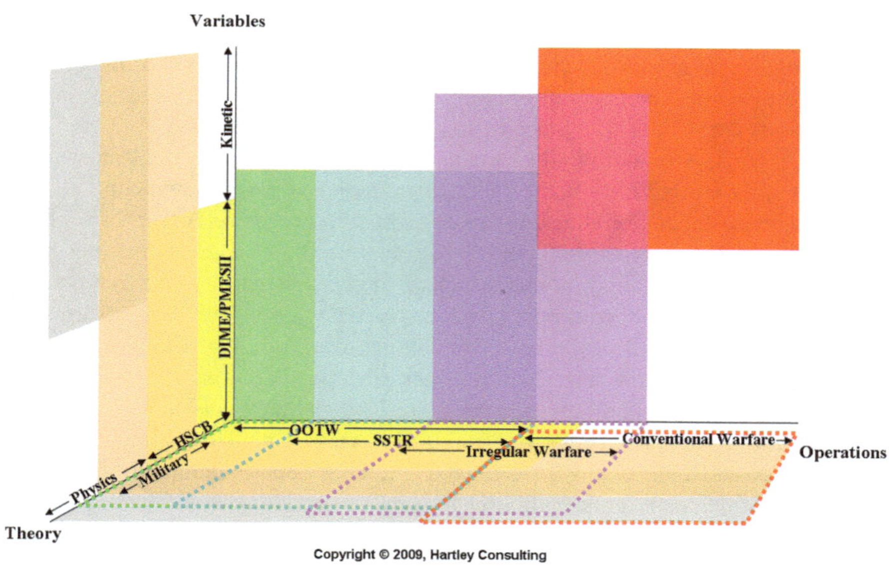

Fig. 1.2 Conflict dimensions

Variables

The vertical axis relates to the variables that may be used to describe the situation. In this figure, they are divided into kinetic and DIME/PMESII variables.

- Kinetic variables include most of the standard variables used in describing conventional warfare, such as probability of kill, the physical characteristics of vehicles, the numbers of personnel and equipment, logistics descriptors, and physical environment descriptors.
- DIME/PMESII variables will be described in more detail in Chapter 4. The important point here is that their focus is on things that are important in OOTW and IW, which are generally missed in the kinetic modeling world. DIME variables are Diplomatic, Information, Military, and Economic actions that can influence the situation. PMESII variables are Political, Military, Economic, Social, Information, and Infrastructure state variables that describe the state of the situation.

Theories

The third axis displays the categories of theories that might be used to explain or predict the results of actions.

- The category labeled "physics" includes all that is often called the hard sciences and mathematics. The theories in this category are generally testable, with fairly well-defined domains of applicability and known degrees of accuracy.
- The category labeled "military," which overlaps the other two categories, includes the various fields taught as military science, which include some hard science and some "softer" subjects, such as military history and leadership.
- The category labeled "HSCB," which stands for Human, Social, Cultural Behavior, contains the fields of social science. The theories in this category are the least well-understood and, unfortunately, the most important for explaining or predicting affairs in unconventional conflict.

Three-Dimensional Volumes

Each of the operational types has rectangles projected onto the (variables × operations) plane and rectangles projected onto the (operations × theories) plane. The combination of these rectangles induces a three-dimensional solid in the conflict space. Unconventional conflict could be considered roughly as the union of the IW solid and the OOTW solid.

Each of the theory categories has rectangles projected onto the (variables × theories) plane and rectangles projected onto the (operations × theories) plane. The induced solids reflect the domains of applicability of the theory category.

The intersections of the two kinds of solids are based on the nature of the theories and the operations. For example, the HSCB solid intersects the conventional war solid in its lower, left, back corner. This indicates that even conventional war cannot be completely described in terms of physical theories.

Modeling Preliminaries

Models are abstractions of reality. That means that parts of what is known about reality are omitted. Ideally, only the parts that are unimportant for the model's purposes are omitted and all of those parts are omitted. These omissions permit the model to be smaller and thus more tractable. The hope is that the resulting model is small enough to make conspicuous the desired features. The decisions on what is unimportant depend on the level of knowledge concerning the domain of interest. Unfortunately, unconventional conflict is a domain in which we are largely ignorant. Further, even with the size reduction accomplished through abstractions, the complexity of the domain results in model sizes that are not very tractable when reasonable fidelity is needed.

Software Design

Early software design practices used the waterfall methodology, which specified identifying the requirements before creating the design, then doing the coding, followed by testing/VV&A and maintenance during use. Figure 1.3 illustrates the waterfall methodology.

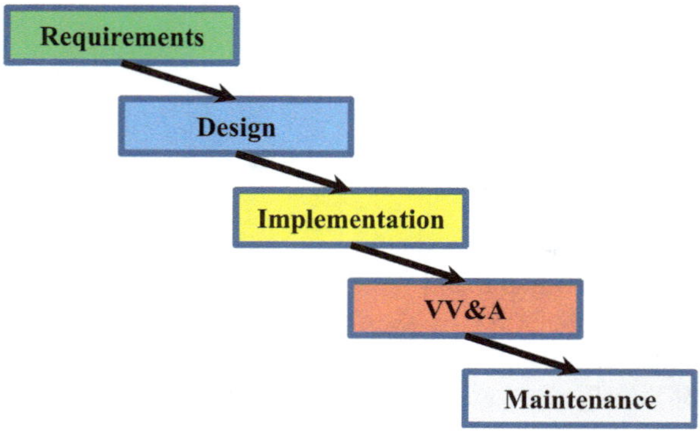

Fig. 1.3 Waterfall software design

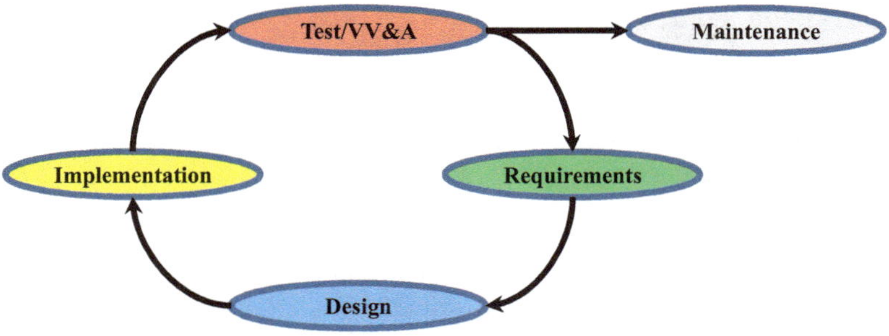

Fig. 1.4 Cyclic software design

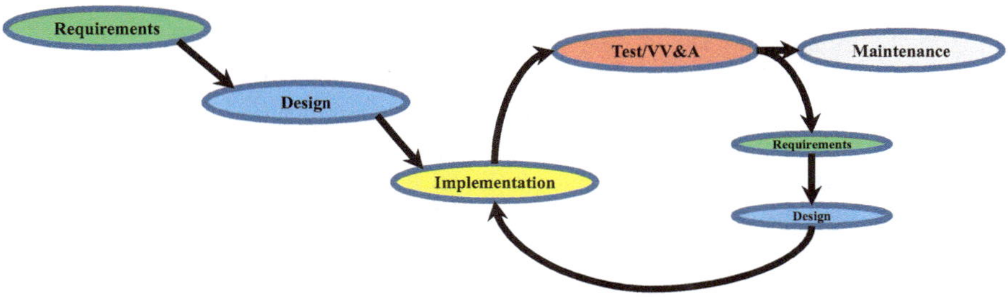

Fig. 1.5 Ontology-based software design

Later practices employ a cyclic process with incremental additions of requirements at the beginning of each cycle (Fig. 1.4), often pictured as a spiral to emphasize the increments. Both of these processes introduce a structure into what would otherwise be a chaotic process of adding "stuff" to the model as it occurs to the developers.

The first cyclic process (several variants have been defined) was probably created from observation. Frequently the requirements can be only partially defined at the start of the process because results along the way feed back into the requirements. Further, it was easier to build a little/test a little and then repeat, rather than to try to build it all, and then test it all.

This was the case for modeling unconventional conflict until an ontology (defined in Chapter 5) for the domain was created. Now, it is possible to define the majority of the requirements at the beginning of the project and to do the majority of the design after that. As a practical matter, it will still be wise to have cycles of requirements, design, coding, and testing (part of verification and validation); however, the requirements and design portions can be smaller parts of the cycle. This is illustrated in Fig. 1.5.

Combat-Type Modeling

The following discussion is taken from the author's contribution to *The Encyclopedia of Operations Research and Management Science* on "Battle Modeling" (Hartley, 2001). We cannot build the ideal battle model; however, we have built many individual battle models, each conceived to fulfill a particular set of objectives. These models of combat may be categorized by their positions along several dimensions; however, they all have one feature in common, and that is the object that is modeled is some aspect of combat. These dimensions with illustrative examples of positions along the dimension are listed as follows:

- **Domain**: land; air; naval; space; combinations
- **Span**: size of conflict—platoon battle; division combat; theater-level combat; global combat
- **Scope**: type of conflict—politico-military; special operations; low-intensity conflict; urban warfare; information warfare; conventional warfare; theater-level nuclear, chemical and biological conflict; strategic nuclear conflict
- **Scoring**: adjudication topics and methodology; measures of merit—attrition, movement, tons of bombs dropped, supplies delivered, victory; methodologies—weapon weights (simple or complex, as in anti-potential potential, which uses eigenvalues to value a weapon by the value of the weapons it can kill), process simulations
- **Randomness**: deterministic or stochastic calculations
- **Combat Activities and Force Composition**: military assets and mission areas—small arms; armor; aircraft; artillery; engineer; logistics; signal; command and control; intelligence; surface navy; submarine; electronic warfare; space assets; missiles
- **Level of Resolution or Detail**: smallest item modeled as a separate entity—bullet; soldier; tank; platoon; company; battalion; brigade; division; corps
- **Environment**: one-dimensional terrain (piston model); two-dimensional terrain (including ocean or air), latitude–longitude or hexagonal grid based; three-dimensional terrain; weather; day–night; smoke
- **Purpose**: design purpose or users' purpose—training; weapon system employment; force composition decisions; operations plans testing
- **Level of Training**: training audience (if any); individual skills; platoon leaders' skills; division staff skills; commanders' skills; combinations
- **Model Treatment of Time**: linear code with no time representation or algorithmically computed time (generally analytic combat models); time-stepped simulations; event-driven simulations; expected value models; stochastic simulations
- **Human Interaction**: data preparation and output interpretation; interruptible with modification and restart; computer-assisted human participation on one or more sides; continuous human participation on all sides
- **Sidedness**: one-sided (e.g., strategic nuclear strike damage effects); two-sided; multi-sided; hard-coded identical properties for each side; hard-coded different

properties for each side (e.g., U.S. vs. Soviet-style tactics); data-driven properties for each side
- **Computer Involvement**: none; moderate; complete
- **Size Computer Required**: PC; mini-computer; mainframe; supercomputer; peripheral equipment required; large run times, small run times
- **External Interactions**: interfaces with parts of the real world—e.g., none; distributed processing; interfaces with weapon simulators; interfaces with real equipment; sand tables; scripting.

Unconventional Conflict Modeling

The dimensions listed for combat modeling also apply to unconventional conflict modeling. The following discussion is modified from the author's contribution to *The Encyclopedia of Operations Research and Management Science* on "Military Operations Other Than War" (Hartley, 2001).

The analytical requirements are characterized by the questions that must be answered. The questions fall into five groups, those that

- Are non-mission related (e.g., what force structure, equipment and plans are needed for the future?),
- Support a decision to engage (or not to engage) in a mission (e.g., what impacts will an OOTW have on other operations? how much will it cost?),
- Are needed to plan a mission (e.g., what is the right force structure? what transport support will we provide to reporters, NGO/PVOs, etc.?),
- Occur during a mission (e.g., which course of action will most quickly accomplish the mission?), and
- Are related to the termination of a mission (e.g., how do we define success and what are its Measures of Effectiveness (MOEs)?).

The question groups are identical to the question groups for combat analysis. Most of the individual questions are also identical. In general, the analysis techniques required to answer the questions are the same. The problem lies in the application: standard applications make assumptions that are valid for combat analysis and invalid for unconventional conflict analysis.

The question of force structure for a mission provides a simple example of the difference between combat analysis and unconventional conflict analysis. For a combat mission, combat troops and equipment are determined first, and the balance of the force structure is composed of the troops and materiel required to support them. Analysis procedures and tools are structured to support this situation. For unconventional conflicts, however, the primary forces may be engineers for disaster reconstruction, medical personnel for disease control, some other "support" function, or combat troops, **depending on the particulars of the mission**. The implied force structure consists of the troops and materiel to support these forces and may (or may not) include combat troops to protect them. Not only are combat analysis procedures and tools set up backwards for unconventional conflict analysis but also unconventional conflict analysis involves multiple possible permutations, requiring significantly more flexibility.

Generally, the characteristics of desirable tools have included simple (e.g., menu driven and point and click), deployable, joint (multi-Service), rigorous, use nonparochial data (not supporting an agenda); **have** available data; are capable of rapid turnaround; and have decision support tools. Analysis tools have ranged from complex simulations of political, economic, sociological, and military interactions to database tools, to spreadsheets, to checklists, with the emphasis on small tools.

Modeling Gaps: Needs versus Availability

Whether the domain is titled OOTW, IW, or unconventional conflict, there is a gap between what needs to be modeled and our ability to model that thing well. Jay Persons reported on the results of a combined U.S. Army/Marine Corps task group that was tasked to conduct a gap analysis of IW methods, modeling, and analysis (Persons, 2010). This task force identified 160 key issues/questions and derived 14 decision issues and 56 required analytic capabilities. It also identified 35 gaps, 20 of which were associated with the social sciences.

These 20 social science gaps required knowledge (theory), data, and algorithms (modeling methods) to model the following:

- The effects of influencers (operational activities) on the attitude/behavior of the civilian population based on ethnic, tribal, cultural, religious, and political considerations;
- The translation of civilian attitudes into levels of cooperation with friendly forces and thus result in corresponding levels of human intelligence (HUMINT) provided by the civilian population;
- The discrimination between civilian and adversary actors based on presented physical and behavioral signatures (e.g., insurgent in civilian clothing);
- The change of affiliations and support for other actors based on the application of influencers (e.g., friendly operations, government activities, and adversary operations);
- The changes in target audience attitudes caused by the application of psychological operations (PSYOPS);
- Adversary HUMINT networks (e.g., attributes of the HUMINT network, how the network is formed, how the network adjusts if a node or element is removed, and what adversary activities tend to facilitate or discourage the population's provision of HUMINT);
- Unique adversary PSYOPS techniques and the effects of those techniques on the target audience;
- The effects of Civil-Military Operations on the attitudes of the civilian population (or other target audience);

- The effects of essential services (or lack thereof) on civilian population's attitudes/behaviors;
- The effects of governmental corruption on governmental institutions and on the civilian population's attitudes/behaviors;
- The attitudes/behaviors of actors (e.g., civilian population) based on the state of physical infrastructure used by the actors;
- Nonhomogeneous groups of actors, particularly problematic when group members have overlapping affiliations (e.g., a single actor or group belongs to multiple groups—religious groups, ethnic groups, political groups, tribal groups, etc.);
- The levels of civilian support for and the provision of physical or monetary resources to adversaries;
- The performance/effectiveness of actor organizations based on the level of cooperation between those organizations (e.g., USAID and local government);
- The behavior of actors (e.g., civilian population and religious leaders) based on their level of support for the existing government;
- The legitimacy of the existing government as viewed from outside the nation by external groups and the effect of international legitimacy on government effectiveness;
- The state of the existing legal system and its impact on the attitudes/behaviors of the civilian population;
- Civilian attitudes/behaviors based on existing economic conditions and how the attitudes/behaviors change as the economic conditions change;
- The effects of media activity on the attitudes/behaviors of actors; and
- The effects of friendly operations on media themes and activity.

Given the creation/discovery of the necessary knowledge/theories and data, the algorithms and modeling methods can be (relatively) easily created. However, until valid theories and data are available, the models that include the behaviors described in these gaps will, of necessity, have lower validity than would otherwise be the case and will need mitigating actions to address these validity problems.

This book does not solve these problems. However, the modeling process that is described in Chapter 7 highlights these hard problems and separates them from the parts that the process makes more straightforward.

Chapter 2
Descriptions of Unconventional Conflict

Before modeling takes place, understanding should be required.

In this chapter, we expand on the definition of unconventional conflict given in the introduction (Chapter 1) by enumerating the many types of operations that are contained in this domain, presenting the results of some relevant analyses, and citing numerous works on the various parts of the domain. The earlier accounts of various types of operations cited in this chapter are more frequently purely descriptive. Later accounts incorporate analysis and some modeling, indicating a growth in understanding.

Unconventional conflict is complex because of the number of relevant actors, the number of actions that can be undertaken, the importance of the passive environment, and the number of relevant metrics. Figure 2.1 illustrates this complexity. It is important to understand that while facts are important, the opinions or sentiment of the actors are also important. Lt. General Flynn (last official position, Director of the Defense Intelligence Agency (DIA)) underlined the criticality of opinions of the populace in obtaining intelligence and for winning in Iraq (Flynn & Ledeen, 2016).

A generic situation will have a Host Nation in which the situation occurs, with its own government, police, and military. The intervening external coalition brings its forces and contractors. Frequently, there are other contractors and non-governmental organizations (NGOs). There may be internal troublemakers (insurgents, terrorists, etc.), and there may be an unfriendly external nation. Each group has its own agenda, with goals and tasks aimed at accomplishing the goals. Also, all actions play out in the same time and space.

The situation can be likened to multiple players playing different board games at the same time on the same board, all interacting, as in Fig. 2.2.

Two players are playing checkers, a game with simple rules and simple strategy. One player is playing chess, a game with more complex rules and more complex strategies. Two players are playing Go, a game with simple rules and complex strategies, and—seven players (blue circled pieces) are playing Monopoly, a game

D.S. Hartley III, *Unconventional Conflict*, Understanding Complex Systems,
DOI 10.1007/978-3-319-51935-7_2

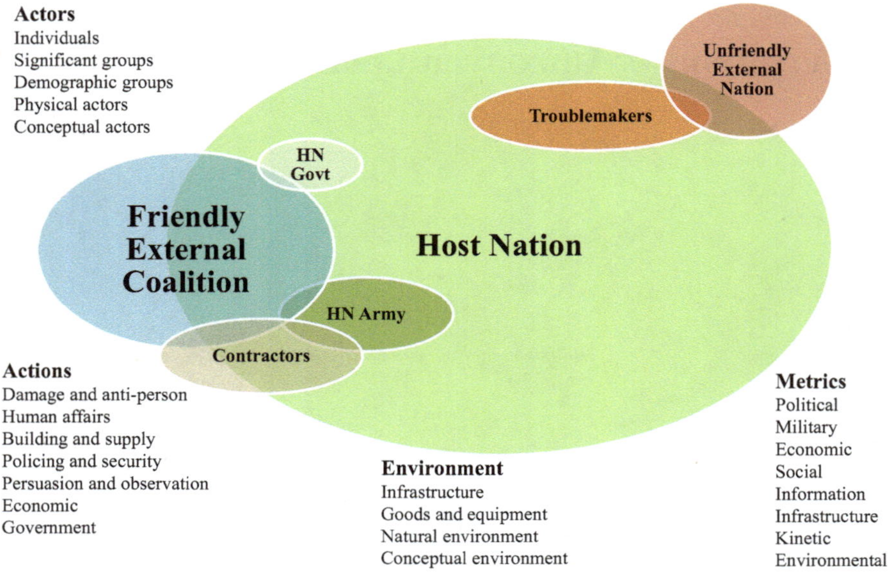

Fig. 2.1 The generic unconventional conflict situation

with complex rules and complex strategies. The combined game has multiple players with unknown rules and very complex strategies.

One of the principal geo-political areas for unconventional conflict is within a failed or failing state. Therefore, it is of great interest to identify such states and predicting which states might fail would be extremely valuable. Section 2 on analyses of unconventional conflict includes work aimed at this.

Types of Operations

Unconventional conflict is also difficult to grasp because of the large number of operations that can be undertaken. It is important to remember that the following definitions are for types of operations, not types of situations. A single situation might require several types of operations, for example, a noncombatant evacuation might be required during a peacekeeping operation that also involves humanitarian assistance. On the other hand, external operations may consist of a single type in the midst of a larger situation, such as a noncombatant evacuation conducted to extract civilians during an insurgency. Despite this difference between operations and situations, an understanding of the possible operations helps in understanding the elements of possible situations.

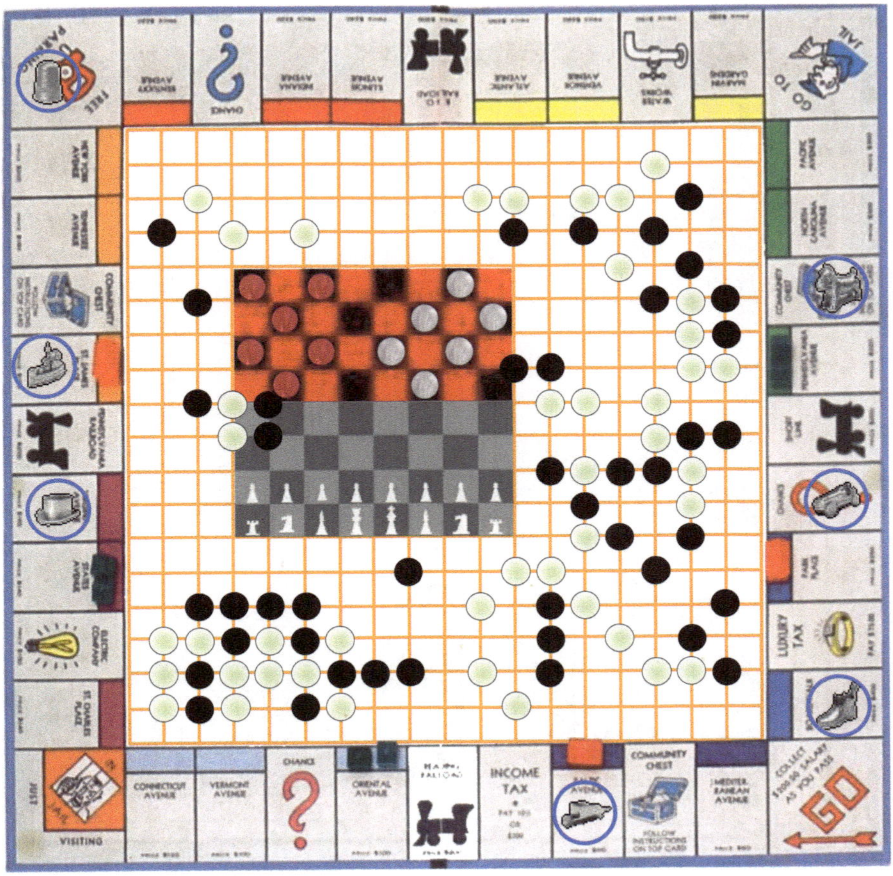

Fig. 2.2 Multiple players, multiple games

Most of these definitions are taken from the report on research done for USPACOM (Hartley, 1996b). These definitions are not authoritative, as different authorities have differing definitions for some of the terms. (See Joint Pub 3-07 for some of the U.S. definitions (Chairman of the Joint Chiefs of Staff, 1995).) However, they are all similar enough that these definitions carry the spirit of the concepts. Citations are included to represent reporting on the particular types of operations.

Two taxonomy diagrams are useful in seeing some of the relationships among these types of operations: OOTW (Fig. 2.3) and unconventional war (Fig. 2.4, following the definitions of the OOTW operations).

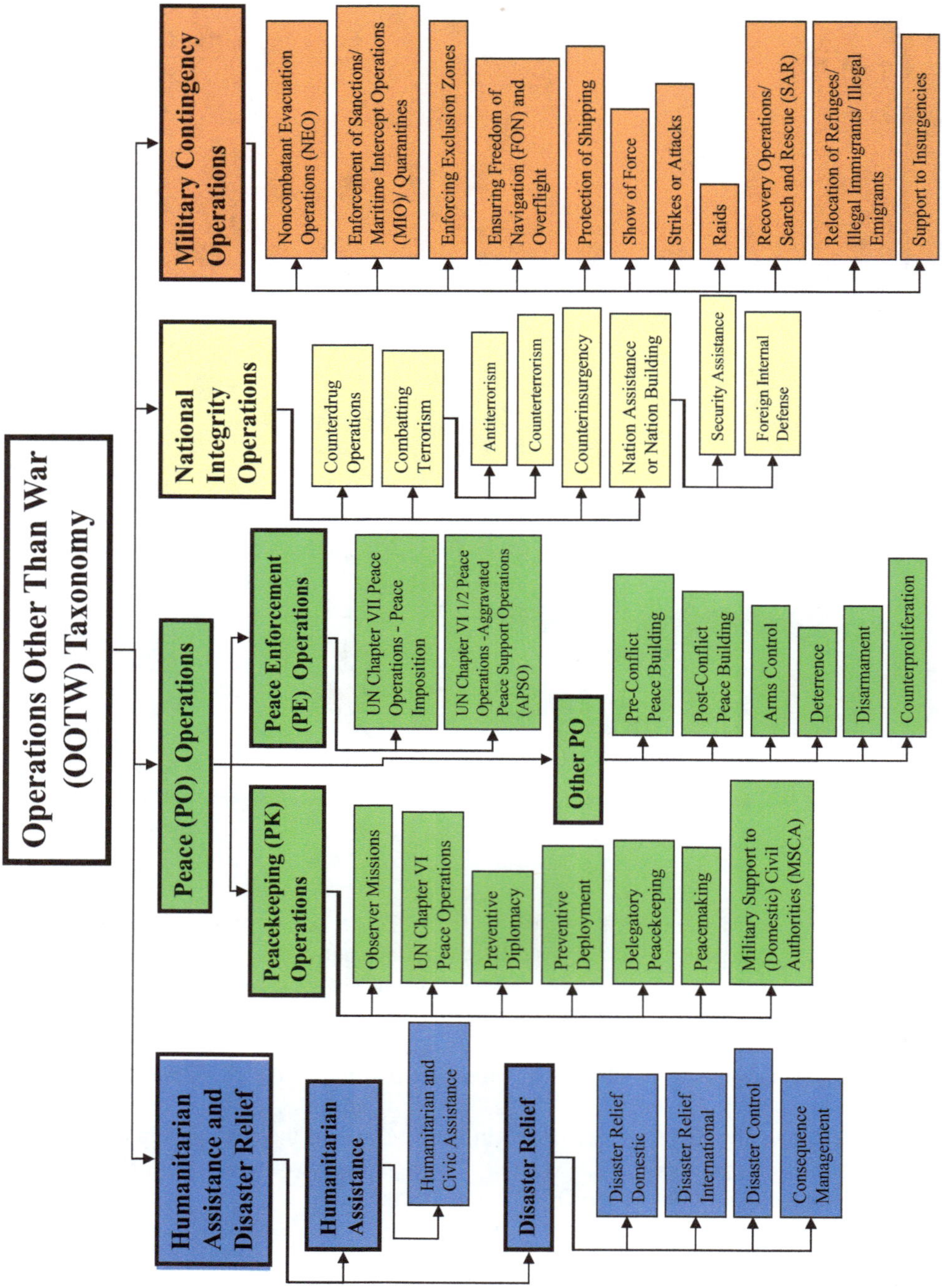

Fig. 2.3 OOTW operations type taxonomy

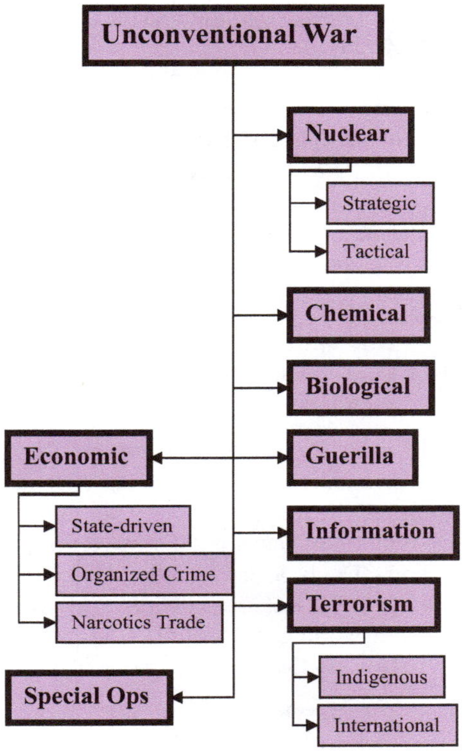

Fig. 2.4 Unconventional war type taxonomy

Humanitarian Assistance and Disaster Relief (HA/DR)

Definition: Missions to promote human welfare, to reduce pain and suffering, to prevent loss of life or destruction of property in the aftermath of natural or man-made disasters. HA/DR includes refugee problems.

In 1995 and 1996, Burkle wrote several useful articles on complex humanitarian emergencies (Burkle, 1995a, 1995b, 1996).

- **Humanitarian Assistance (HA)**: Definition: Programs conducted to relieve or reduce the results of natural or man-made disasters or other endemic conditions such as human pain, disease, hunger, or privation that might present a serious threat to life or that can result in great damage to or loss of property. In those operations in which governmental structures have broken down, the military may be in charge through the Civil-Military Operations Center (CMOC) to provide essential humanitarian and technical expertise with the goal to contain the situation and transition to another lead agency.

  In 1993, Carol Clair wrote about humanitarian assistance (Clair, 1993) and Jonathan Dworken wrote about relations with humanitarian relief organizations (Dworken, 1993).

In 1994, Svijac and McGrady wrote about Operation Restore Hope (Zvijac & McGrady, 1994).

In 1995, Adam Siegel created a report on humanitarian assistance and peace operations (Siegel, 1995) and Katherine McGrady and others wrote about a humanitarian assistance game (McGrady et al., 1995).

In 1996, Sandra Newett wrote about planning for humanitarian assistance operations (Newett, 1996), Mark Geis wrote about logistics and engineering planning for humanitarian assistance operations (Geis, 1996), Jonathan Dworken wrote about improving Marine coordination with relief organizations (Dworken, 1996), and Newett, et al. wrote about Emerald Express '95 (Newett et al., 1996).

- **Humanitarian and Civic Assistance**: Definition: Incidental assistance to the local populace provided in conjunction with military operations and exercises.

• **Disaster Relief (DR)**: Definition: Disaster relief falls within the overall context of humanitarian assistance but is conducted in emergency situations to prevent loss of life and property.

In 1990, Davis and Farnsworth wrote about hurricane disaster relief operations (Davis & Farnsworth, 1990).

In 1992, Kate Farnsworth, et al. wrote about disaster relief in Southern Africa (Farnsworth et al., 1992).

In 1994, Steven Metz wrote about disaster in Rwanda (Metz, 1994).

- **Disaster Relief, Domestic**: Definition: U.S. disaster relief operations; the Federal Emergency Management Agency (FEMA) is in charge; the military is an asset to local and state governments bringing specific capabilities to contain the situation and assist in recovery.
- **Disaster Relief, International**: Definition: The host government is intact and requests assistance to handle a natural or man-made crisis; the military is an asset to local governments or international agencies to bring specific capabilities to contain situation and assist in recovery.
- **Disaster Control**: Definition: Measures taken before, during, or after hostile action or natural or man-made disasters to reduce the probability of damage, minimize its effects, and initiate recovery.
- **Consequence Management**: Definition: Measures taken after a Weapons of Mass Destruction (WMD) attack to alleviate the damage, loss, hardship, or suffering; restore essential government services; protect public health and safety; and provide emergency relief to affected governments, businesses, and individuals. FEMA is the designated lead agency for U.S. domestic operations.

Peace Operations (PO)

Definition: Military operations to support diplomatic efforts to reach a long-term political settlement (includes both peacekeeping and peace enforcement).

In 1995, Alberts and Hayes produced a book about command arrangements in peace operations (Alberts & Hayes, 1995) and Adam Siegel created a report on humanitarian assistance and peace operations (Siegel, 1995).

In 1996, Hayes and Wheatley wrote about the interagency and political-military dimensions of peace operations (Hayes & Wheatley, 1996).

In 1998, Pascale Combelles-Siegel described information activities in peace operations (Combelles-Siegel, 1998).

In 1999, Avruch, Narel, and Combelles-Siegel wrote about information campaigns in peace operations (Avruch, Narel, & Combelles-Siegel, 1999).

- **Peacekeeping Operations (PK)**: Definition: Military operations undertaken with the consent of **all** major parties to a dispute, designed to monitor and facilitate implementation of an agreement and support diplomatic efforts to reach a long-term political settlement. Often involves ambiguous situations requiring the peacekeeping force to deal with extreme tension and violence without becoming a participant. This also known as **UN Chapter VI Peace Operations**.

  In 1992, the General Accounting Office (GAO) produced a report on U.S. peacekeeping operations (General Accounting Office, 1992).

  In 1993, Durch edited a report on UN peacekeeping operations (Durch, 1993).

  In 1995, Daniel Shedlowski discussed peacekeeping operations (Shedlowski, 1995) and the Dupuy Institute issued a report on peacekeeping in Bosnia (The Dupuy Institute, 1995).

  In 1996, Rose and Lambert discussed operations in Bosnia (Rose & Lambert, 1996).

  In 1997, Larry Wentz compiled a book on lessons from peacekeeping in Bosnia (Wentz, 1997).

  - **Observer Missions**: Definition: A type of peacekeeping operation—assisting in the observance and maintenance of a cease-fire; acting as a neutral witness for the handing-over of personnel or property from one party to another; and other limited operations.
  - **Preventive Diplomacy**: Definition: A type of peacekeeping operation—diplomatic actions taken in advance of a predictable crisis to prevent or limit violence.
  - **Preventive Deployment**: Definition: A subset of peacekeeping (consent is assumed) in which (military) forces are deployed prior to hostilities erupting with a goal of preventing active conflict.
  - **Delegatory Peacekeeping**: Definition: Peacekeeping operations led by regional organizations but sanctioned by the UN.
  - **Peacemaking**: Definition: A type of peacekeeping operation—the process of diplomacy, mediation, negotiation, or other forms of peaceful settlements that arrange an end to a disput and resolve issues that led to conflict.
  - **Military Support to (Domestic) Civil Authorities (MSCA)**: Definition: When authorized, armed forces assist in domestic emergencies within the continental U.S.; the Army has primary responsibility. (Under provisions of

the *Posse Comitatus Act,* neither the active component nor the U.S. Army Reserve may execute the law in the place of duly appointed law enforcement means without specific Presidential or Congressional approval and direction.)

- **Peace Enforcement Operations (PE)**: Definition: The authorized application of military force to compel compliance with resolutions or sanctions designed to maintain or restore peace and order.
  - **UN Chapter VII Peace Operations**: Definition: Part of peace enforcement—operations, short of war, requiring force to impose peace—also known as **peace imposition.**
  - **UN Chapter VI ½ Peace Operations**: Definition: Operations falling between UN Chapter VI and UN Chapter VII, and thus often referred to as Chapter VI ½. Operations requiring a show of force, or small tactical operations, to enforce peace—part of peace enforcement, also known as **Aggravated Peace Support Operations (APSO).**
- **Other Peace Operations**:
  - **Preconflict Peace Building**: Definition: Longer-term, non-military, economic, social, and political measures that can help states deal with emerging threats and disputes.
  - **Postconflict Peace Building**: Definition: Postconflict actions, predominantly diplomatic and economic that strengthen and rebuild governmental infrastructure and institutions in order to avoid a relapse into conflict. May be associated with either peacekeeping or peace enforcement.

    In 2011, Paul Davis edited a volume on postconflict stabilization and reconstruction (Davis, 2011). This work rests on the premise that the success of stabilization and reconstruction requires successes at some threshold levels of each of the four components: security, political (governance) stability, social stability, and economic stability.
  - **Arms Control**: Definition: Any plan, arrangement, or process, resting upon explicit or implicit international agreement, governing the numbers, types, and characteristics of weapon systems or the numerical strength, organization, equipment, deployment, or employment of armed forces. Arms control focuses on promoting strategic military stability. Arms control encompasses disarmament. Depending on the situation during implementation, it may be either peacekeeping or peace enforcement.
  - **Deterrence**: Definition: May consist of either actions or maintenance of a particular state, such as level of preparedness, that creates negative incentives for another country or group to engage in war.

    In 2010, Pat McKenna presented a paper on deterrence, focusing on the changes needed to account for the non-state actors in unconventional conflicts (McKenna, 2010). He pointed out that a deterrence study must address three questions: (1) Who is the adversary? (2) What is the decision in question? (3) Under what circumstances is influence desired? In unconventional conflict, the deterrence may need to flow from actions ostensibly directed at

another party, through whom the influence is planned to flow to the potential adversary. However, those same actions will have other effects on other parties, all of which must be accounted for.
  - **Disarmament**: Definition: The reduction of a military establishment to some level set by international agreement. Depending on implementation situation, may be either peacekeeping or peace enforcement.
  - **Counterproliferation**: Definition: Efforts to impede the proliferation of weapons of mass destruction (WMD: chemical, biological, and nuclear or radiological weapons).

National Integrity (NI) Operations

Definition: Operations to promote national integrity.

- **Counterdrug (CD) Operations**: Definition: Support to federal, state, and local law enforcement agencies in their efforts to disrupt the transfer of illegal drugs into the United States. Those active measures are taken to detect, monitor, and counter the production, trafficking, and use of illegal drugs. Support efforts to interdict the flow of illegal drugs at the source, in transit, and during distribution.

  In 1995, Eric Shaw wrote about the Andean drug war (Shaw, 1995).

  In 1997, Coyle and Alexander wrote about the drug trade (Coyle & Alexander, 1997).
- **Combatting Terrorism**: Definition: Actions taken to oppose terrorism from wherever the threat.

  In 1994, Bruce Hoffman wrote about terrorism (Hoffman, 1994).

  In 1995, P. K. Rosbolt wrote about Islamic fundamentalism (Rosbolt, 1995).

  In 2009, Davis and Cragin edited a volume on social science for counterterrorism (Davis & Cragin, 2009). This book includes chapters on the causes of terrorism, the reasons people have for joining, the support mechanisms for terrorism, the economics of terrorism, the organizational decision making, and the end processes for terrorist groups.

  In 2015, Durante and Fox wrote about finding the root causes for violent acts in insurgencies and terrorism in the Philippines (Durante & Fox, 2015). They found that the three main contributors to significant increases in the number of violent acts in the Philippines were poverty levels (positive), illiteracy (positive), and satisfaction with the government (negative).

  In 2016, Lt. General Flynn defined four tasks for defeating radical Islamists in his book, *The Field of Fight*: destroy the jihadi armies and their leaders, discredit their ideology, create a new set of allies, and challenge the regimes that support the radical Islamists (Flynn & Ledeen, 2016).

  In 2016, Sebastian Gorka wrote a book about defeating the Jihadist movement (including ISIS, Al Qaeda, and others). He included a thorough discussion of the history, thought processes, and agendas of the significant groups and the movement as a whole. He also included a brief discussion of the appropriate agenda for the United States, including three strategic tasks: (1) Deploy the truth (explain the true nature of the enemy in a massive counterpropaganda campaign; strengthen the counterpropaganda efforts of our allies and partners); (2) Help others fight their own wars (employ foreign internal defense operations; secondarily, use our forces to locate and kill the worst of the worst in the jihadist movement); (3) Win the war at home (educate and train the armed services, the federal, state, and local police forces, and the intelligence community in the enemy threat doctrine; employ the New York Police Department's model of counterterrorism intelligence across the federal government) (Gorka, 2016).

- **Antiterrorism**: Definition: Part of combatting terrorism—those passive defensive measures taken to minimize vulnerability to terrorism.
- **Counterterrorism (CT)**: Definition: Part of combatting terrorism—the full range of offensive measures taken to prevent, deter, and respond to terrorism. Counterterrorism occurs in both unconventional conflict and war.

- **Counterinsurgency (CI or COIN)**: Definition: Those military, paramilitary, political, economic, psychological, and civic actions taken by a government to defeat insurgency. Or, the use of military resources to provide support to a host nation's counterinsurgency operations in the context of Foreign Internal Defense (FID) through logistical and training support.

In 1993, Steven Metz wrote about the future of insurgency (Metz, 1993) and David Hogg wrote about gangs (Hogg, 1993).

In 1994, Blanco wrote about counterinsurgency in Cuba (Blanco, 1994).

In 1995, P. K. Rosbolt wrote about Islamic fundamentalism (Rosbolt, 1995).

In 1996, Coyle and Millar wrote about counterinsurgency in Rhodesia (Coyle & Millar, 1996).

In 2010, John Sokolowski discussed a study of an insurgency in Colombia (Sokolowski, 2010). The focus of the study was on how policy changes in Colombia might affect insurgency strength. The systems model used had 125 factors for Colombia, of which Sokolowski believed 60 could apply outside of Colombia.

In 2010, Jack Goldstone presented a paper on modeling counterinsurgency operations (Goldstone, 2010). Goldstone made the distinction between a *tracking* model that shows how a situation is responding to actions and a *forecasting* model that "aims to determine how changes in the actions of various actors or shifts in government policies will affect the insurgency." He said tracking models are difficult to create but far easier than forecasting models, which require causal modeling.

Goldstone then discussed what an insurgency is and the need to balance political and security objectives. After this, he discussed technical issues for tracking models and forecasting models. Finally he discussed harnessing tracking models to forecasting models to support planning and recalibrating counterinsurgency operations.

In 2015, Durante and Fox wrote about finding the root causes for violent acts in insurgencies and terrorism in the Philippines (Durante & Fox, 2015). They found that the three main contributors to significant increases in the number of violent acts in the Philippines were poverty levels (positive), illiteracy (positive), and satisfaction with the government (negative).

In 2016, Unal discussed the coevolution of strategies and tactics of the Turkish Army and Kurdish insurgents over more than 30 years of counterinsurgency operations (Unal, 2016). King, Hering, and Aguilar wrote about building predictive models of counterinsurgent deaths (King, Hering, & Aguilar, 2016). They used robust clustering analyses to categorize the data and regression techniques to make the predictions. King, Hering, and Newman wrote about schemes for classifying counterinsurgencies for use in comparing historical counterinsurgencies to current counterinsurgencies (King, Hering, & Newman, 2014). They evaluate three classical classification schemes: Insurgent Strategy, "which groups insurgencies by the military methods insurgents used to fight;" Insurgent Type, "which groups insurgencies based on the ideology of the insurgent;" and Actor-Centric classification, "which groups insurgencies using the counterinsurgent's mission." They ask "Which Coin classification scheme is best for estimating the outcome of a counterinsurgency?" They conclude that the Insurgent Strategy classification scheme yields the best results.

See also the discussion on FM 3-24 Counterinsurgency in Chapter 3, Models.

- **Nation Assistance or Nation Building**: Definition: U.S. support of a host nation's efforts to promote development, ideally through the use of host nation resources.

  In 1991, Robin Cababa wrote about nation assistance (Cababa, 1991).

  - **Security Assistance**: Definition: Providing defense material, military training, and defense-related services by grant, loan, credit, or cash sales to further U.S. national policies and objectives. May take place in either nation building or counterinsurgency operations.
  - **Foreign Internal Defense (FID)**: Definition: Programs that encompass the total political, economic, informational, and military support provided to another nation to assist its fight against subversion and insurgency. FID also includes participation by civilian and military agencies of one government in any of the action programs taken by another government to free and protect its society from subversion, lawlessness, and insurgency. It may take place in either nation building or counterinsurgency operations.
  - **Stability Operations**: Definition: Various military missions, tasks, and activities conducted outside the United States in coordination with other instruments of national power to maintain or reestablish a safe and secure environment, provide essential governmental services, provide emergency infrastructure reconstruction, and provide humanitarian relief.

    The Army Field Manual 3-07 provides the U.S. Army view of Stability Operations (HQDA, 2008b).

    In 1995, Berg and Hamby created a general work on regional stability assessment (Berg & Hamby, 1995) and John Cowan wrote about defending Kurds fleeing their homes in northern Iraq following the Gulf War (Cowan, 1995).

In 2010, Richard Deckro discussed issues in Stability, Security, Transition, and Reconstruction Operations (SSTRO) (Deckro, 2010). He noted the DoD commitment to its role in SSTRO and the fact that the Secretary of State is the lead in stabilization and reconstruction activities. The Department of State (DOS) does not have the manpower or skills to plan for these operations.

In 2011, Paul Davis edited a volume on stabilization and reconstruction (Davis, 2011). This work rests on the premise that the success of stabilization and reconstruction requires successes at some threshold levels of each of the four components: security, political (governance) stability, social stability, and economic stability.

Military Contingency Operations

Definition: Military contingency operations are very similar to traditional military operations. To the extent that traditional operations are understood and adequate tools created, military contingency operations analysis requirements are met. However, there are holes in the tools for traditional operations and some specialized models may be needed. An example is transportation planning for opposed NEOs.

- **Noncombatant Evacuation Operations (NEO), Opposed**: Definition: Operations to relocate noncombatants from a foreign country where the relocation is opposed with armed force.
- **Noncombatant Evacuation Operations (NEO), Unopposed**: Definition: Operations to relocate threatened noncombatants from a foreign country or host nation that are not opposed by armed force. It may involve threatened U.S. citizens.

  In 1992, David Stahl wrote about noncombatant evacuation operations (Stahl, 1992).

  In 1995, Ray Clark also wrote about noncombatant evacuation operations (Clark, 1995).

  In 2008, Dell and Sparling presented a paper on optimal distribution of resources for noncombatant evacuation operations (Dell & Sparling, 2008). This classified paper won the Rist Prize and is reported to have generated interest in its application.
- **Enforcement of Sanctions/Maritime Intercept Operations (MIO)**: Definition: A type of military contingency operation involving coercive measures to interdict the movement of certain types of designated items into or out of a nation or specified area.
- **Enforcing Exclusion Zones**: Definition: A type of military contingency operation to prohibit specified activities in a specific geographic area.
- **Ensuring Freedom of Navigation (FON) and Overflight**: Definition: Operations conducted to demonstrate U.S. or international rights to navigate sea or air routes.

- **Protection of Shipping**: Definition: U.S. forces providing protection of U.S. flag vessels, U.S. citizens, and their property against unlawful violence (such as piracy) in and over international waters (Fellman, 2015).
- **Show of Force Operations**: Definition: A type of military contingency operation carried out to demonstrate U.S. resolve in which U.S. forces deploy to defuse a situation that may be detrimental to U.S. interests or national objectives. Can take the form of combined training exercises, rehearsals, forward deployment of military forces, or introduction and buildup of military forces in a region.
- **Strikes or Attacks**: Definition: Offensive operations conducted to inflict damage on, seize, or destroy an objective for political purposes or to demonstrate U.S. capability and resolve to achieve a favorable result.
- **Raids**: Definition: Usually a small-scale operation involving swift penetration of hostile territory to secure information, confuse the enemy, temporarily seize an objective, or destroy installations.
- **Recovery Operations/Search and Rescue (SAR)**: Definition: The search for location, identification, rescue, and return of personnel or human remains, sensitive equipment, or items critical to national security.
- **Relocation of Refugees/Illegal Immigrants/Illegal Emigrants**: Definition: Transporting, and often caring for, refugees or other detained persons.
- **Support to Insurgency**: Definition: Support to an organized movement aimed at the overthrow of a constituted government through the use of subversion and armed conflict.

Unconventional Warfare

Definition: Unconventional warfare is a catch-all category for warfare that does not follow mid-twentieth century rules. Some discussions of this from a U.S. point of view are found in Joint Pub 3-05.1 (Chairman of the Joint Chiefs of Staff, 2001).

- **Radiological Warfare**: Definition: The use of radioactive materials with the intent to damage an organization or nation by killing or incapacitating humans, animals, or plants or by damaging property.
- **Chemical Warfare**: Definition: The use of toxic chemicals (including biological toxins) with the intent to damage an organization or nation by killing or incapacitating humans, animals, or plants or by damaging property.
- **Biological Warfare**: Definition: The use of biological toxins or infectious agents, such as bacteria, viruses, and fungi, with the intent to damage an organization or nation by killing or incapacitating humans, animals, or plants.
- **Economic Warfare**: Definition: The use of any means of which the primary effect is to damage the economy of an opponent. It can involve physical actions such as blockades, economic actions such as freezing bank accounts, social actions such as supporting organized crime or narcotics trade, etc.

In 1993, David Hogg wrote about gangs (Hogg, 1993).

In 1995, Eric Shaw wrote about the Andean drug war (Shaw, 1995).

- **Guerrilla Warfare**: Definition: Warfare in which a small group of combatants such as paramilitary personnel, armed civilians, or irregulars use military tactics including ambushes, sabotage, raids, petty warfare, hit-and-run tactics, and mobility to fight a larger and less-mobile traditional military.

  The 1969 Army FM 31-21 (HQDA, 1969) defined guerrilla warfare as consisting "of military and paramilitary operations conducted in enemy held or hostile territory by irregular, predominantly indigenous forces. The term guerrilla warfare is used to denote the primary overt military activities of a resistance movement," and said guerrilla warfare is "characterized by offensive actions with emphasis on swift, brief, violent action, and elusiveness, mobility, and surprise. In addition, there are other important considerations such as local population support, external support, and the political, legal, and tactical situation."
- **Information Warfare and Information Operations**: Definition: The use and management of information and communications technology in pursuit of a competitive advantage over an opponent. This includes operations directed at military and economic infrastructure and those operations directed at the opinions of populations. It includes propaganda and counterpropaganda and cyberwar, attacks, and defenses.

  In 1998, Pascale Combelles-Siegel described information activities in peace operations (Combelles-Siegel, 1998).

  In 1999, Avruch, Narel, and Combelles-Siegel wrote about information campaigns in peace operations (Avruch et al., 1999).
- **Terrorism**: Definition: The use of violence, or threatened use of violence, to achieve a political, religious, or ideological aim. It is considered a war crime under the laws of war when used to target noncombatants, such as civilians, neutral military personnel, or enemy prisoners of war.

  In 2010, Will Moore analyzed terrorism in the context of the behavior of governments and dissidents, rather than focusing on polities, economies, and societies (Moore, 2010). Jerrold Post discussed a typology of terrorism, with political terrorism, criminal terrorism, and pathological terrorism as the roots (Post, 2010). He expanded political terrorism to include religious extremist terrorism and included descriptions of the motivation and psychology of terrorists, and Lailari Guermantes discussed problems in terrorism and counterterrorism models (Guermantes, 2010).

  In 2010, Dipak Gupta presented a paper on the dynamics of terrorist movements (Gupta, 2010). He said, "[d]espite popular belief that poverty, income inequality, and lack of political freedom cause terrorism, the econometric studies have come up largely empty handed. The reason for this apparent lack of correlation rests with the fact that while the structural variables offer the necessary condition for political violence, the sufficient cause rests with the ability of a political entrepreneur to take aspects of frustration and frame the grievances in terms of a matter of collective identity, complete with a plan of action."

Gupta discussed the details behind these statements and then discussed increasing and decreasing forces that affect the dynamics of terrorism. He ended with a discussion of the dynamics of terrorist movements—the interactions between the dissident group and the state authorities and the argument that accepting the need for a catalytic agent prevents prediction of the rise of terrorist groups.

In 2013, Shmuel Bar discussed radicalization of Muslims, the leading cause of current terrorist activity (Bar, 2013). He argued "the key drivers to radicalization in Muslim societies are therefore not an individual rebellious response to dissatisfaction in society by rejection of any authority, but rather a replacement of the 'compromising' authority of parents, teachers, and Imams with alternative—militant, uncompromising, and seemingly pristinely Islamic authority. In this sense, the Muslim radical is, in fact, a conformist."

In 2015, Fellman, Bar-Yam, and Minai published a book on terrorism, etc. (Fellman, Bar-Yam, & Minai, 2015).

In 2016, Neil Johnson et al. examined the online presence of supporters of ISIS and discovered an "ecology evolving on a daily time scale that drives online support" and provided a mathematical theory describing this evolution (Johnson et al., 2016). They also made a prediction "that development of large, potentially potent pro-ISIS aggregates can be thwarted by targeting smaller ones."

- **Special Operations**: Definition: Military operations that are "special" or unconventional and carried out by dedicated Special Forces units using unconventional methods and resources. These include drone operations.
- **Low Intensity Conflict (LIC)**: Definition: LIC was coined to describe operations like those early in the Vietnam conflict in which small unit combat took place occasionally, interspersed with periods of tense waiting and maneuvering.

  In 1994, Kenneth Allard wrote about operations in Somalia. This work is now available in the commercial press (Allard, 2002).

Analyses of Unconventional Conflict

Understanding a domain not only requires definitions but also requires investigations, analyses, and attempts at modeling.

1994 and 1995 USPACOM Beginnings

In 1994, in an INFORMS panel discussion, Dave Haut, then Chief of the Research and Analysis Division for the Department of Defense's U.S. Pacific Command (USPACOM) asked the question, "Suppose there are problems in country *X* and the Ambassador has the choice of having a carrier battle group sail down the country's coastline in a show of force or playing golf with the country's Prime Minister; how does he decide which will be more effective?" (Haut, 1994) The combat models of the time had no way of framing such a situation, much less any hope of answering

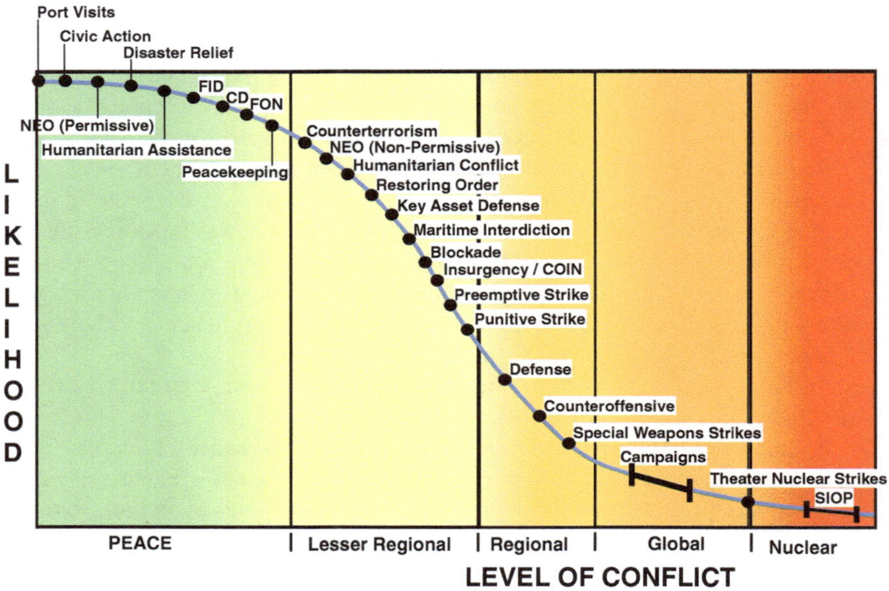

Fig. 2.5 Haut's continuum of operations

the question. We still are unable to answer this question; however, we are getting closer.

In a presentation to the Japan-US Operations Research Seminar (JUORS) in 1995, Haut presented an estimate of the likelihood of various types of operations across the continuum from peace to nuclear war (Haut, 1995). Figure 2.5 displays this figure, which shows that the various types of OOTW operations are much more likely to occur than are conventional or nuclear combat operations.

1996–1999 Workshops and Projects on OOTW

USPACOM initiated a project on OOTW tool requirements (Hartley, 1996b). This project involved a set of three workshops attended by operators (the people who have to carry out the operations), analysts (the people who analyze situations and plan operations), academics, and modelers. The project also involved document research and attending externally sponsored conferences.

One of the initial products was a graph of the number of OOTWs by type that the United States supported over several years. The graph (Fig. 2.6) shows from 20 to 40 OOTWs per year in the years from 1990 through 1996. The "CONT" in the legend refers to military contingency operations. These values provide evidence of the general correctness of the likelihood estimates in Haut's figure.

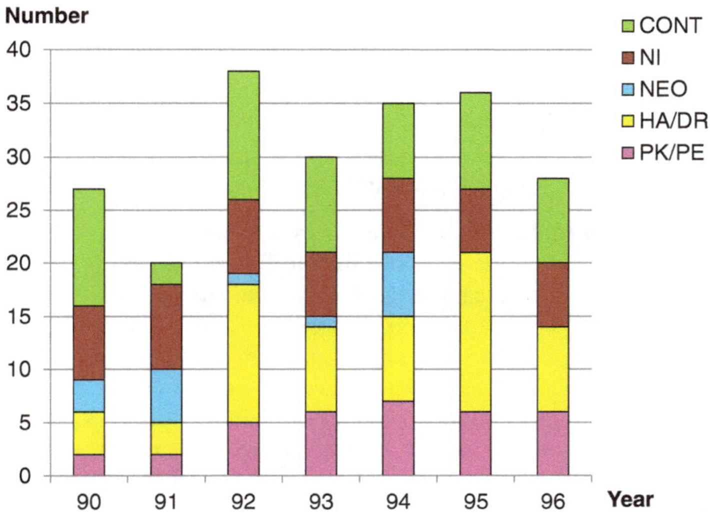

Fig. 2.6 U.S. support to OOTWs by type and year

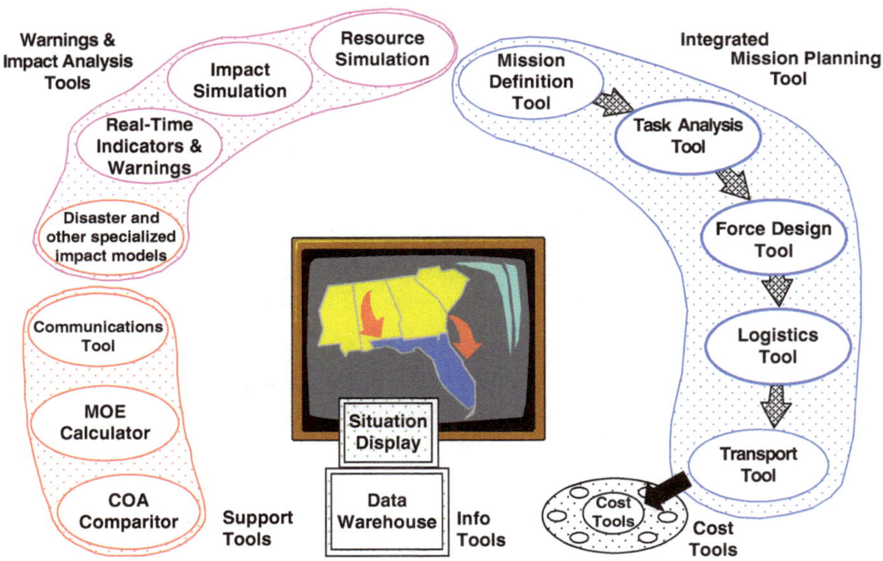

Fig. 2.7 Required tool categories to support OOTW

The purpose of the USPACOM project was to review what was known about OOTW and define the tool categories that were needed to support the various operations that were being conducted and those that were envisioned in the future. It should be noted that some of these needs still remain unfilled. Figure 2.7 displays the results of this project graphically. The following verbal description of the results gives the essentials for each of the tool categories. In each case where "OOTW" is used, "unconventional conflict" can be substituted.

- **Warnings and Impact Analysis Tools**: These tools are among the most difficult (scientifically) to create but are essential to the analysis of OOTWs. Three tools are included in this group as follows:
  - The **real-time indicators and warnings tool** serves to filter, interpret, and track world news in the light of possible future OOTWs: several attempts were being made to create tools of this type, such as the Protocol for the Assessment of Nonviolent Direct Action (PANDA) (Bond & Vogele, 1995).
  - The **impact simulation** models the significant relationships included in and surrounding an OOTW to permit prediction of the results of actions, whether human or environmental: the commercial computer game, SimCity™ 2000, (SimCity 2000, 1994), is an example of an impact simulation. Unfortunately, the nature of social interactions is a matter for debate and consequently the proper mathematical expressions of these interactions and the best methods for modeling them are undecided. While at least two candidate simulations existed at the time, Spectrum (National Simulation Center, 1996) and the Deployable Exercise Support system/Civil Affairs Module (DEXES/CAM) (Woodcock, 1996), these were regarded with some misgivings by working analysts, apparently because of lack of transparency or because they were used for training. The Situational Influence Assessment Module (SIAM) (Hayes & Sands, 1997) used another technique to address social interactions. It was an influence diagram-based model, not a simulation model, but could be useful in this category.
  - The **resource simulation** models the changes in resource consumption and sequestration (allocation to some other use) over the course of an OOTW: this need could be satisfied by the Joint Warfare Simulation (JWARS) (Hartley & Packard, 1998b), which was being created at the time.
- **Integrated Mission Planning Tool**: The five separate tools that comprise this group should ultimately be seamlessly integrated, although the initial integration may be loose. Each tool feeds its successor, while permitting reentry for iterative planning. These tools are relatively simple (scientifically); however, to be useful in an OOTW context, they require careful definition with respect to applicability to joint, coalition (multi-country), and non-military component analysis (a definite lack at the time). The tools are a mission definition tool, a task analysis tool, a force design tool, a logistics tool, and a transportation tool.
  - The **mission definition tool** should provide a "reality check" to ensure that all implications of the mission are fully understood. The Conceptual Model of Peace Operations (CMPO), a peace operations influence diagram-based checklist, was an example (Davis, 1996).
  - The object of the **task analysis tool** is to support an accurate and complete analysis of the mission tasks. The tool needed is a decision support tool that

connects missions to strategies to tasks, both explicit and implied, in the OOTW domain. It should identify both those tasks that are central to the mission and any contingent tasks that might be implied by reasonable shifts in mission definition. It should also support replanning as the situation changes. Lidy produced the data to support such a tool (Lidy, 1998).

- The object of the **force design tool** is to support the designation of U.S. forces required for an operation in an OOTW context. The tool needed is a decision support tool that connects the tasks to generic resources and connects generic resources to actual available resources, including U.S. military, U.S. non-military, foreign government, NGO/PVO, and contractor resources. Data requirements include task capability for all resources (or the facility for user input of unique resources) and availability data (based on reserve commitments, etc.). It should provide for restrictions on choices based on cultural issues. Processing should include selection of military resources and substitution of other resources. The tool should also support replanning as the situation changes (Hartley, Bell, & Packard, 1999).
- The object of the **logistics analysis tool** is to support the logistics analysis of the mission in an OOTW context. The tool needed is a decision support tool that derives the logistics requirements from the total force structure. It should allow for supply from outside sources and provide for supply of non-military personnel. It should support replanning as the situation changes. In 1998, Brundage et al. investigated the availability and utility of existing tools of this type (Brundage et al., 1998).
- The object of the **transport analysis tool** is to support the transportation analysis for mission arrival, sustainment, and departure in an OOTW context. The tool needed is a decision support tool that plans the transport requirements, based on all appropriate constraints. It must support replanning when the situation changes after some transport has been accomplished. The Joint Flow and Analysis System for Transportation (JFAST) and the Model for Intertheater Deployment by Air and Sea (MIDAS) were examples of this type tool.

• **Support Tools**: This group contains three specific tools and a cluster of several tools related by type. The **COA comparator** permits the development of courses of action (COAs) through several levels of alternatives: an influence diagram/ decision tree methodology would support this type analysis. The **MOE calculator** supports the calculation and tracking of MOE values. The **communications tool** supports planning the communications system within the complex context of OOTWs. The cluster of **disaster impact tools** (e.g., hurricanes, volcanos, earthquakes, fires, and nuclear accident) supports the estimate of the situation in several technical areas, such as engineering and health. The Consequence Assessment Tool Set (CATS) supported some of these functions.
• **Cost Models**: Seven tools make up this group. Their object is to calculate the cost information for various aspects of OOTWs: incremental costs of notional OOTWs, to support long-term analysis; probable incremental costs, to support

the decision on engaging in a particular OOTW; relative (full) costs, to support the selection of the mission plan; costs incurred, to support cost recovery from other U.S. agencies and from foreign organizations and governments; incremental costs of a particular OOTW, to support the Congressional Budget process; costs of a particular OOTW, including equipment depreciation, readiness losses, increased reserve recruitment and training costs, and perhaps other costs, to support future acquisition, budgeting, and training decisions; and actual costs of a completed OOTW, to support improved estimates of future operations and reports to Congress on actual costs. Work was initiated to address analysis tools (Institute for Defense Analyses, 1998; Hartley & Packard, 1998a).
- **Information Tools**: There are two tools in this category. The **situation display** presents the information concerning the situation in a manner designed to support visualization and maximize understanding: the Virtual Information Center (VIC) project represented a first attempt at creating this type tool (Sovereign, 1998). The **data warehouse** either stores or provides links to (as appropriate) all pertinent data. The data and their usability are critical to good analysis in the OOTW domain, as well as in the combat domain. However, the data required for OOTW analysis and the display requirements are in an embryonic state when compared to the state of affairs of combat analysis.

1998–1999 OSD, MORS, and SO/LIC Follow-Ons

The Office of the Secretary of Defense (OSD) Program Analysis and Evaluation (PA&E) commissioned a follow-on study to examine the overlap between a large new combat model, the Joint Warfare Simulation (JWARS), and the needs for OOTW tools (Hartley & Packard, 1998b).

Figure 2.8 shows one of the diagrams that expresses the results of the OSD (PA&E) project. The horizontal axis shows mission scale, with example missions outside of the graph: at the low end are Special Operations Forces (SOF) missions, military contingency operations (MCO), and smaller-scale contingency (SSC) operations, and at the high end are Major Theater War (MTW) and global operations. The vertical axis shows areas for military activities: combat; transport, logistics, etc.; data requirements; and political, economic, and social activities (now called DIME/PMESII, without the military part). The blue rectangle shows the approximate range of JWARS capabilities from MTW down to some SOF missions and all but the DIME/PMESII activities. The magenta free-form shape shows the OOTW needs coverage, overlapping a portion of the JWARS rectangle, extending beyond it at the small-scale end, and covering most of the DIME/PMESII region.

JWARS has since gone the way of many other models—shelved; however, the point remains true for other combat models. Any model that is not designed to address DIME/PMESII issues either cannot address them or will do so poorly.

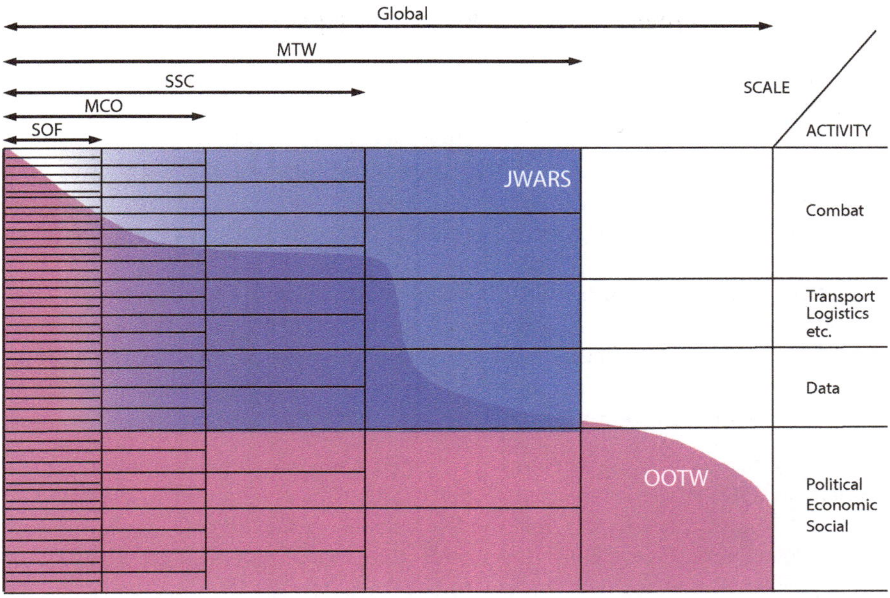

Fig. 2.8 OOTW coverage by JWARS

A parallel and supporting activity involved cochairing a MORS workshop on OOTW analysis and modeling techniques (Staniec & Hartley, 1999).

Following the JWARS project, the Office of the Assistant Secretary of Defense (OASD) Special Operations/Low Intensity Conflict (SO/LIC) created three projects. The first was to examine the cost modeling needed for OOTWs (Hartley & Packard, 1998a). Seven types of cost models were defined as follows:

- **Incremental costs of notional OOTWs**, to support the long-term analysis: the concept is to have a simple cost model of a generic Noncombatant Evacuation Operation (NEO), a generic Freedom of Navigation (FON) operation, etc., each with only a few variables, representing cost drivers.
- **Probable incremental costs**, to support the decision on engaging in a particular OOTW: the concept is a complete cost model that permits iterative refinements as the mission becomes increasingly well defined.
- **Relative (full) costs**, to support the selection of the mission plan: the concept is for the model to support comparison of the costs of alternative COAs, whether incremental or full costs are required for valid comparisons.
- **Costs incurred**, to support recovery of those costs from other U.S. agencies and from foreign organizations and governments: the concept is to ensure that all recoverable costs are identified, supporting differences in allowable cost definitions, depending on the organization paying the bill.
- **Incremental costs of a particular OOTW**, to support the Congressional Budget process: the concept is to ensure that long-term, ongoing OOTWs (e.g., the

United Nations Sinai Peacekeeping operation) and foreseeably repetitive OOTWs (e.g., South American counterdrug operations) are correctly budgeted.

- **Cumulative costs of a particular OOTW**, including equipment depreciation, readiness losses, increased reserve recruitment and training costs, and perhaps other costs, to support future acquisition, budgeting, and training decisions: the concept is to capture the costs of replacing capabilities lost through participation in OOTWs, for example, excess depreciation caused by excessive operational tempo (OPTEMPO) and loss of personnel in reserve units caused by frequent call-ups.
- **Actual costs of a completed OOTW**, to support improved estimates of future operations and reports to Congress on actual costs: the concept is to capture actual costs in accounts that can be compared to those of the forecasts, permitting improved forecasts of ongoing operations and better planning factors to generate improved future forecasts.

The second SO/LIC project involved specifying the force definition model needed for OOTWs (Hartley et al., 1999). This project required a workshop of operators, analysts, academics, and modelers to ensure that all considerations were taken into account. Figure 2.9 shows the resulting model specification. The process looks similar to that for defining the forces for a combat mission; however, the important distinction is that the "critical units" might be such things as water purification units, rather than combat units, and the support units might be combat units, rather than the typical support units.

The third SO/LIC project was to support a USPACOM workshop to investigate modeling techniques for predicting national instabilities (Hartley, 1999). Five models were described in detail and rated as the best prospects as follows:

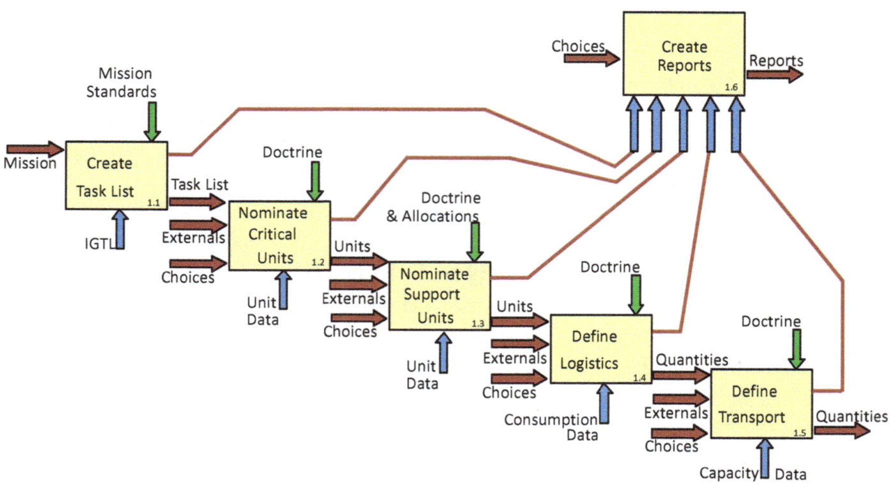

Fig. 2.9 Force model specification overview

Table 2.1 Instability prediction model techniques

	ACT	EBR	SFP	PRISM	JWAC
Artificial neural networks	X		X		
Chaos theory				X	
Fuzzy analysis	X				
Game theory					X
General systems theory					X
Genetic algorithms					X
Hidden Markov methods					X
Linguistic analysis					X
Logistic regression	X	X	X		
Rule induction	X				
Social network analysis					X
System stability analysis					X

- The Analysis of Complex Threats (ACT) is the Center for Army Analysis methodology,
- EBR is the Evidence Based Research, Inc. (EBR) methodology,
- The State Failure Project (SFP) is the Central Intelligence Agency (CIA) methodology,
- PRISM (not an acronym) is the National Ground Intelligence Center (NGIC) methodology, and
- JWAC is the Joint Warfare Analysis Center (JWAC) methodology.

Table 2.1 shows the technical techniques used for each of these five methodologies to predict national instability.

Other Analyses

Later work, described by David Carment and his colleagues, laid out a framework for country risk analysis and early warning (Carment, Prest, St. Jean, Samy, & Wyjad, 2010). They discussed the definitions, concepts, and strategies of several other groups and some generic techniques, such as country-specific structural data, dynamic data analysis using structured events coding, and expert opinions, surveys, and polling. They then described their approach, the Country Indicators for Foreign Policy (CIFP). They produced a fragility index by combining several analyses.

- They described and measured three essential properties of a state:
  - Authority,
  - Legitimacy, and
  - Capacity.

- They measured one cross-cutting theme: Gender.
- They then measured several indicator clusters:
  - Governance,
  - Economics,
  - Security and Crime,
  - Human Development,
  - Demography, and
  - Environment.

In addition to the static fragility index, they employed events monitoring using coded events related to the indicator clusters and scored as stabilizing or destabilizing. The combination gave a fragility index and a vector of future change.

Ferleman and Walters used a different methodology for forecasting instability to examine the state of the Middle East in 2040 (Ferleman & Walters, 2010). They based their methodology on five independent variables as follows:

- Democracy,
- Infant mortality relative to the global average,
- Trade openness as indicated by exports plus imports as a percentage of GDP,
- GDP per capita purchasing power parity, and
- The average number of years of education of the population at least 25 years old.

Richard Lobban took an empirical approach in examining the Sudan as a classic case of a failed state (Lobban, 2010). Important variables included the following:

- Cultural,
- Historical,
- Ecological,
- Religious,
- Political, and
- Racio-ethnic variables.

Other significant factors included the following:

- The nature and identity of foreign intervenors,
- The rebel movements,
- Janjaweed,
- Sudanese political players, and
- External-to-Sudan issues and political players.

There have been many other attempts at understanding unconventional conflict through analyses. Some of these have involved building testable models, such as the economic models of Berman and Kuznar (Berman & Kuznar, 2010). One such model was the "club" model, which applied to religious organizations, sects, and mafias. The idea was that the members cooperate to produce some local public good (such as law and order) for their own benefit. They developed internal relationships that are resistant and resilient to defection.

In 2013, Cassani et al. discussed a research effort involving population sentiment analysis (Cassani et al., 2013a). This involved multidimensional analysis of attitudinal survey data and a Bayesian analysis of the data to determine influence patterns within the data.

In 2013, Yohai et al. discussed combining social and environmental models (Yohai et al., 2013a). They "demonstrate how two climate events—severe rainfall that hinders a road construction project, and drought that causes poor crop harvests—can both influence opinions toward various actors, such as coalition forces, the Afghan government, and the Taliban, and response planning based on these anticipated opinion changes."

Chapter 3
Modeling Research

Trial and error is better than no trial; it may lead to learning.

Before the U.S. government started worrying about unconventional conflict, science fiction authors had posited models of human behavior. The most popular early entry was Asimov's psychohistory (Asimov, 1951). The concept was an analog of the gas laws of physics: although the motion of a particular molecule of gas is essentially random, when large enough numbers of gas molecules are present, the actions of the mass of gas molecules can be predicted. Similarly, laws of human actions in the aggregate should be discoverable. Michael Flynn discussed how this might become reality (Flynn, 1988).

In this chapter, we will look at several modeling constructs that have been advanced as organizing principles for models of unconventional conflict. We then review the many technical modeling approaches that might be used for these models. We follow with the descriptions of several models and systems of models that have been created to answer questions about unconventional conflict. Then we conclude with the results of a series of workshops concerning modeling requirements.

As shown in Fig. 2.6 above, the United States has been constantly involved in unconventional conflicts at least since 1990 and has a past history of involvement in them. Because the U.S. military provides the bulk of the manpower, equipment and funding for these operations, it is interested in figuring out how to do them better. This has resulted in research into the subject, including research into modeling the domain.

The U.S. Department of Defense (DoD) thrives on acronyms: we have already seen OOTW, SSTR, and IW. DoD also creates new concepts with new names and acronyms regularly. It has been accused of doing this to confuse people outside of the inner circle; however, the truth lies more with the fact that conflict is so complex that no one truly understands it now, nor did anyone in the past. Therefore, when someone comes up with a new way of looking at some facet of the problem, a new name is required and, of course, a new acronym.

D.S. Hartley III, *Unconventional Conflict*, Understanding Complex Systems,
DOI 10.1007/978-3-319-51935-7_3

One of the first attempts to create a comprehensive understanding of modeling unconventional conflict involved the search for analysis tools to support OOTW (Hartley, 1996b).

Modeling Constructs

Several different conceptual constructs have been created to address unconventional conflict.

Psycho-Social Attributes

Before there was IW and before there was OOTW, there was Low Intensity Conflict (LIC), coined to describe things like the early Vietnam operations. In 1995, ten countries held the Asia-Pacific Military Operations Research Symposium III (AMORS III) in Thailand to discuss peace and LIC (AMORS III, 1995).

One of the papers (Hartley, 1995) was later expanded (Hartley, 1996a). One key concept was the introduction of psycho-social factors into a military model. As shown in Fig. 3.1, the transitions from one model state to another would be driven by the values of the psycho-social factors.

Segmentation by Agent/Object

A second key concept in this paper was the introduction of different types of agents, other than the purely military hierarchy of traditional military models. As shown in Fig. 3.2, in addition to country-sized agents, selected important

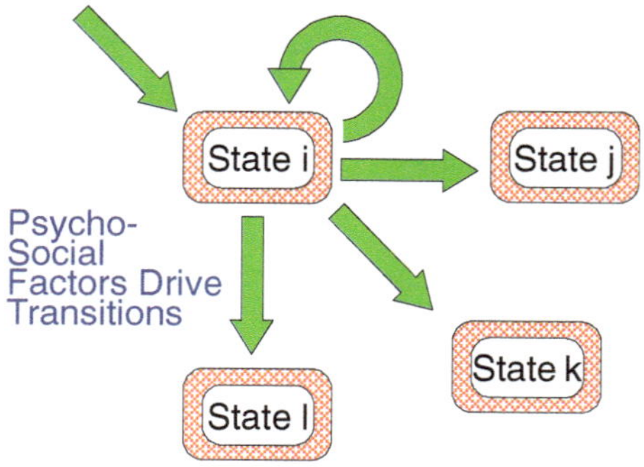

Fig. 3.1 Psycho-social factors and state transitions

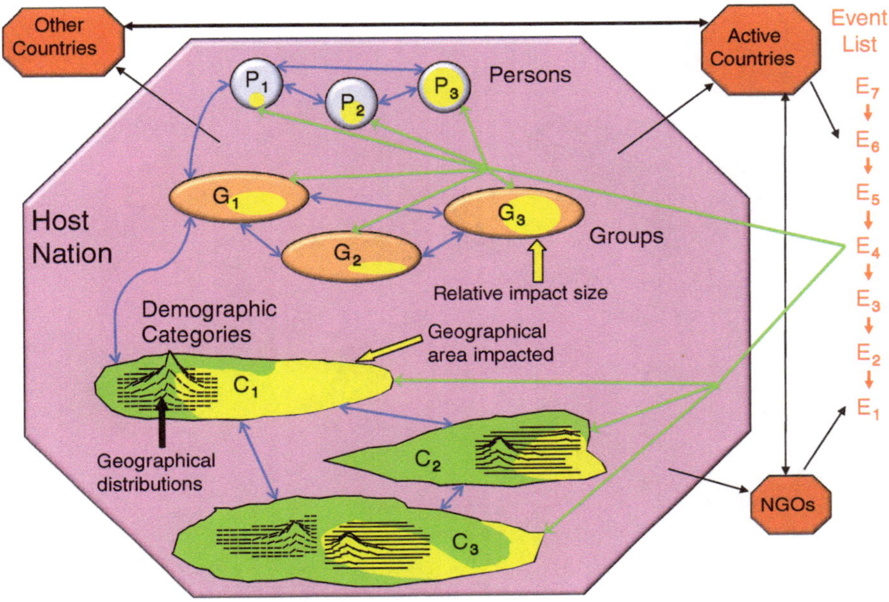

Fig. 3.2 Agents and events

individual persons, selected important groups, and demographic categories were introduced as agents.

The third key concept was the introduction of an event list to control the execution of the model. Agents could place events on the event list. As an event is executed, its effects on all of the agents would be calculated.

Segmentation by Factors

One of the first models relating to unconventional conflict was created through a series of workshops (Hayes & Sands, 1997). These workshops were part of a government-sponsored project to examine how military operations could support the long-term objective of creating civil stability and a durable peace in countries involved in unconventional conflict. The participants were experts in the many disciplines and activities relating to "complex emergencies" from various governmental agencies, academic institutions, and nongovernmental organizations.

The model that was created was an influence diagram in essence. Figure 3.3 is a redrawing of the links among the factors defined in the model, grouped to reduce external links among the groups and to maintain a coherence of meaning. The factor groups are the following:

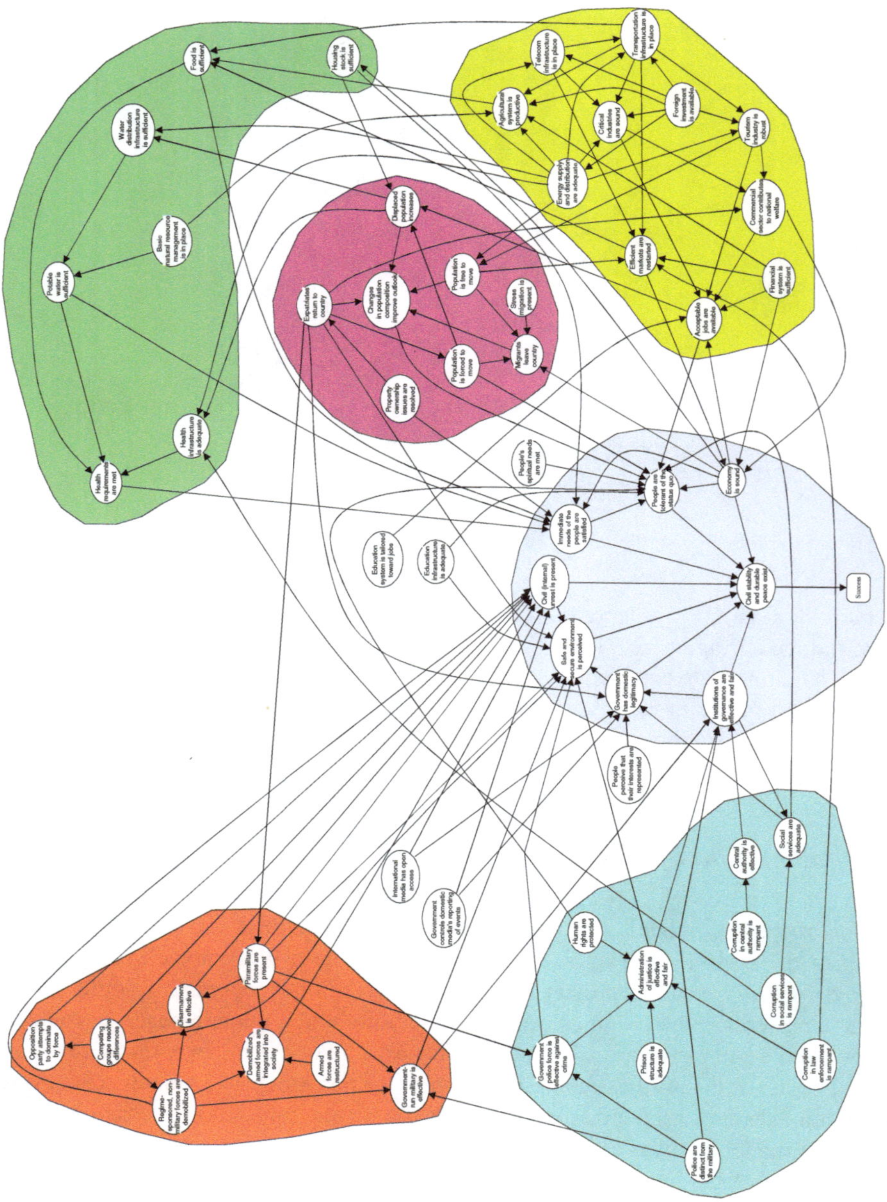

Fig. 3.3 Factor modeling in doing windows

- Armed conflict,
- Immediate needs,
- Population movement,
- Justice system,
- Economy,
- Media,
- Education, and
- Core influences (the final output measures).

The factor groups in Fig. 3.3 are labeled, with the individual factors represented by the small ovals (the contents of which are not meant to be readable). This model represented a new step in modeling in which factors that influence results are given pre-eminence. Accordingly, it is called factor modeling.

A 1998 Military Operations Research Society (MORS) workshop continued the work on tools for OOTW (Staniec & Hartley, 1999). Several potentially useful factors were identified and discussed in this workshop, such as the following:

- Military factors (force structure, rules of engagement (ROEs)),
- Political factors (global, local, domestic),
- Economic factors (local, exterior support),
- Geographic factors (distance, terrain),
- Cultural factors (religious, dietary),
- Mission factors (initial, minor changes, transitions),
- Psychological factors (personality, media),
- Organizational factors (coalitions, NGOs), and
- Interactions among factors.

In 2003, I began a process of putting together a model based on these concepts. I created the Interim Semi-static Stability Model (ISSM) (Hartley, 2006a). Figure 3.4 illustrates the initial logic of the ISSM. It was an influence diagram, instantiated as an Excel workbook. This model implemented the Doing Windows concept as closely as was possible given the differing capabilities of Excel modeling and the SIAM tool used in the original work. The final output was an estimate of the level of civil stability and durable peace that results from the input variable values. The variables are grouped in roughly the same areas as those shown in Fig. 3.3 (again, the individual variables are not meant to be readable).

Later, I added factors that I called Interventions. These interventions would now be called DIME variables and the original variables would be called PMESII variables (Hartley, 2006a). (For those interested in more details, the pdf file of the Users' Guide can be downloaded by using the link to the first reference cited at the bottom of the page http://drdeanhartley.com/HartleyConsulting/TOOLBOX/issmdesc.htm.) The interventions were implemented with an important difference from the input variables of the original model. The values for the original input variables were obtained by observing the situation and estimating the appropriate values. The values for the intervention variables consisted of the effort employed to change outcomes. The new logic consisted of connections among the interventions

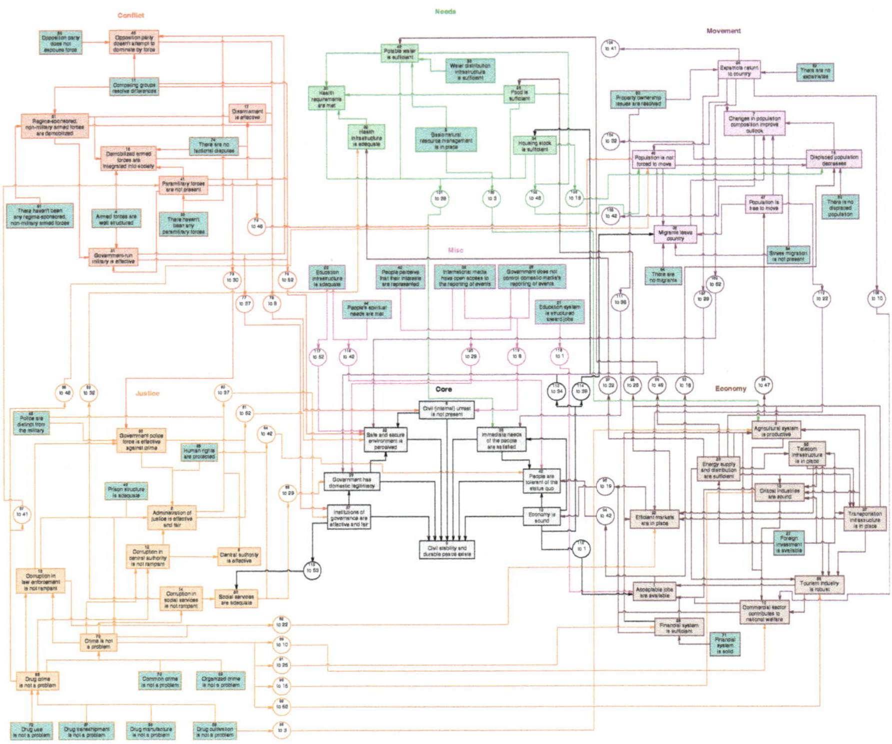

Fig. 3.4 ISSM Version 1 influence diagram

and the variables they would impact, time delays in the flow of changes, and formulas for evaluating the amounts of impact produced. Figure 3.5 illustrates the new logic and shows where the various components reside in the logic.

Segmentation by PMESII Category

By 2006, after the creation of the ISSM, the PMESII concept was solidified. (Chapter 4 examines this concept in detail.) Figure 3.6 shows a U.S. Army construct (HQ DA, 2006b) showing the six PMESII domains, each with its own internal set of links and nodes and links among the domains. The point of this diagram was that the operational center of gravity (COG) could be in one domain (in this figure, the military domain), while the strategic COG could be in another domain (here the political domain). In physics, the center of gravity of an object is the central point of the object with respect to the weights of the constituent particles of the object. In the military, the COG of an organization is the source of strength of the organization.

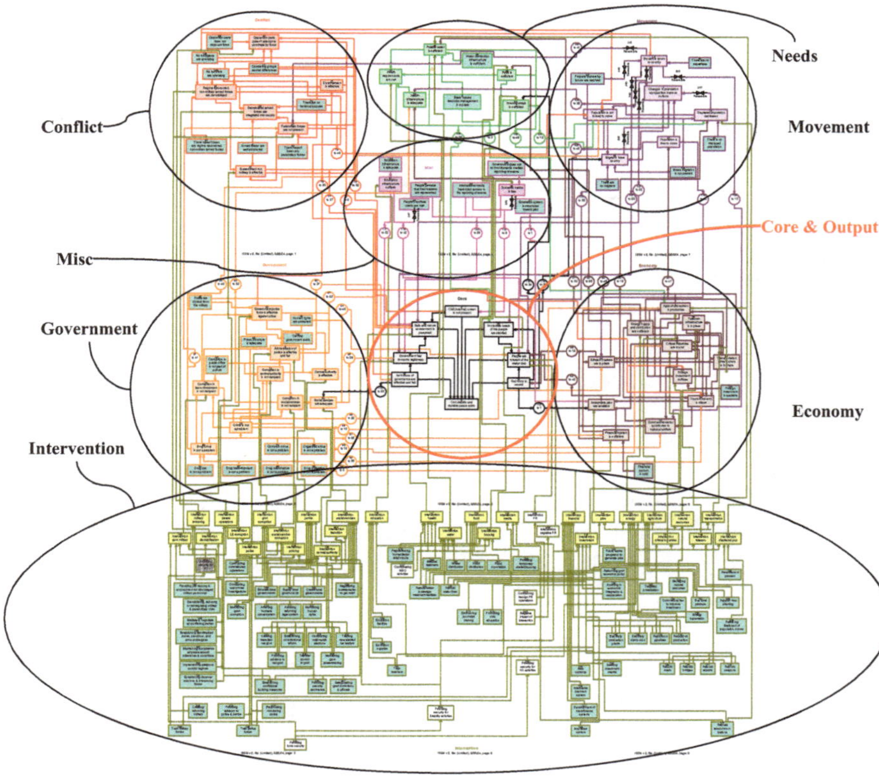

Fig. 3.5 Influence diagram of later ISSM versions

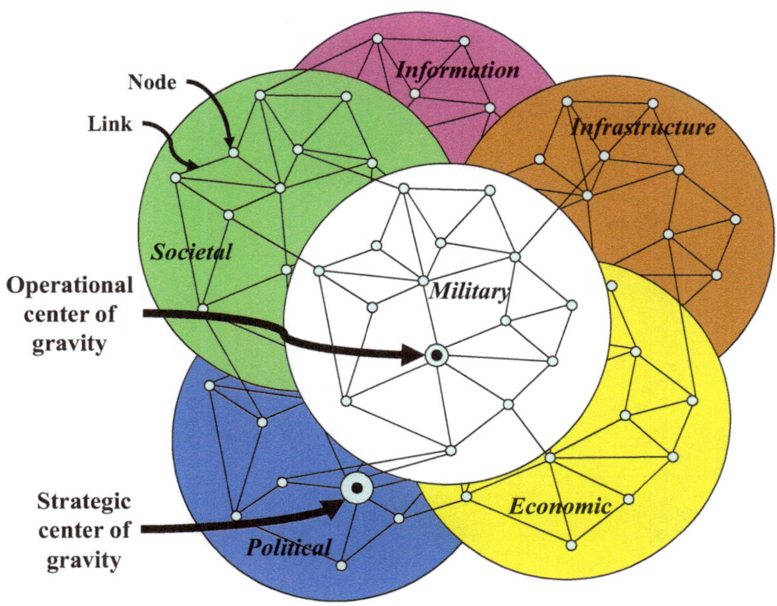

Fig. 3.6 PMESII segmentation

The concept was invented by von Clausewitz to describe the location of the cohesive strength of a military force—the point at which an attack should be aimed (Clausewitz, 1993).

Segmentation into "Battles"

Hartley, Holdsworth and Farrell described a segmentation of the PMESII system into activity types (Hartley, Holdsworth & Farrell, 2006). The HQ DA segmentation was by the PMESII components and the ISSM segmentation was by factors. The Hartley, Holdsworth and Farrell segmentation was labeled "Battles" and was derived from an unknown source within the Department of Defense.

Figure 3.7 shows these "Battles" in an irregular conflict, with their connections. (This is the part from the unknown source.) The diagram may be read by imagining a group of people who want to cause disruption or take power in the region. First

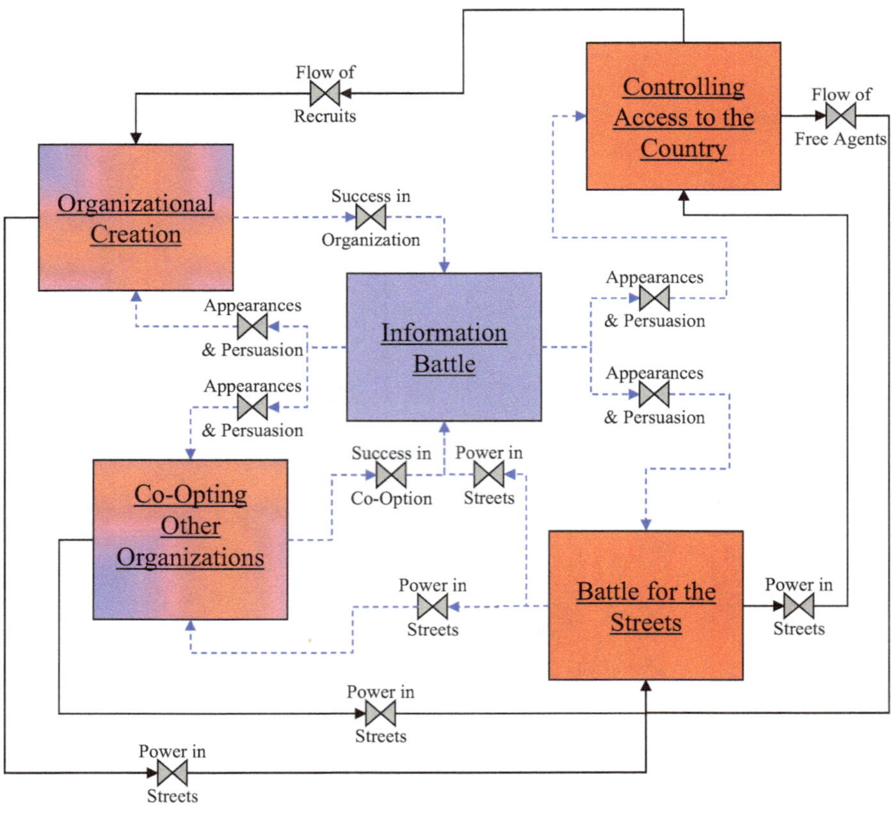

Fig. 3.7 Irregular "Battles"

they must create an organization, which involves an information "battle" and a cycle of appearance and persuasion in which they are successful. Then they want to co-opt other organizations to their way of thought and to pursuing their aims or aims that support them. This also involves an information campaign. At some point their success is great enough to start looking for more control, the "Battle for the Streets," and to gain recruits, internally and externally. Both of these involve information campaigns. As the organization gains success, resistance sets in and there is a "Battle" for control of access to the country at the borders. The links described so far all involve converting popular and individual opinion to their side. The other links represent the physical flows of people (recruits) and power (usually destructive, but also services provided). Naturally, each of these "Battles" is (at least) two-sided, with resistance from the government and from other groups with their own ideas of what should be done. Clearly, no kinetic (attrition and combat damage) model is adequate to the task of measuring all the outputs of these "battles." A DIME/PMESII approach is required.

As complex as the picture in Fig. 3.7 is, it is inadequate. Something critical is missing and that is the effect this activity has on the daily life of the rest of the population—and the effect this daily life has on the "battles" so far defined. Figure 3.8 adds the "Daily Life Battle." This "battle" includes the economy (broadly defined), governmental activities, the basic needs of the populace, and the level of crime. It is the evaluation of "normalcy" that is the critical measure of success.

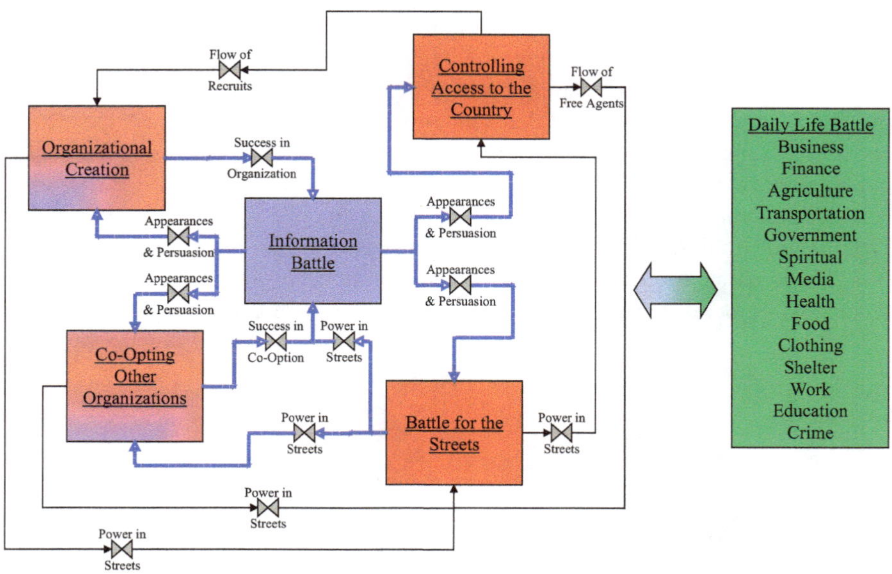

Fig. 3.8 Irregular "Battles" plus daily life

Punctuated Simulation

Figure 3.9 illustrates the punctuated analysis construct, also developed in Hartley, Holdsworth and Farrell. At the beginning of the analysis, the starting assumptions are input into a PMESII tool (shown as an ISSM tracking chart) (Hartley, 2006a). This may be a single start point (white "x" within the icon) or a set of points representing the "road to war," the events leading up to the starting point for the analysis. This process ensures that the results of the study are grounded with a value for the measure of success to compare to the results from the various simulations. It also ensures that the study assumptions are logical with respect to a PMESII view of the world, not just arbitrary assumptions that may represent an unlikely scenario. The analysts then convert this result into the inputs for the tools in use for each of the battles.

Each of the tools within each of the battles is run for a particular simulated period. The tools are stopped and their outputs are used as input to the PMESII tool. This process generates a new measure of success (point on the tracking chart). Converting the outputs of the various models into inputs for the PMESII tool requires that the analysts put all the pieces together and, with the support of the PMESII tool, make sense of the results. In general, the analysts will be required to modify the results of some of the tools because there will be disagreements among them or because they are known to be immature or because they are known to be missing key factors in their makeup.

It is at this point that the conceptual methodology really diverges from previous analysis methodologies. The analysis is a **punctuated simulation**, with human inputs required between pieces of the simulation. The analysts take the results from the PMESII tool and create new starting points for the simulations that begin at the same simulated time as the ending of the previous run. This allows the analysts to implicitly combine the results of the simulations, making up for the inadequacies of each. For example, suppose one tool simulates the economic aspects of the battle, but ignores social factors, and another simulates the social aspects, but ignores economic factors. The analysts use the ending points of each tool as the basis for their new beginning points, but modify them based on the results from the other tool. Suppose that the economic model shows an improving economy and the social model shows increasing factional conflict. The analysts might restart the economic

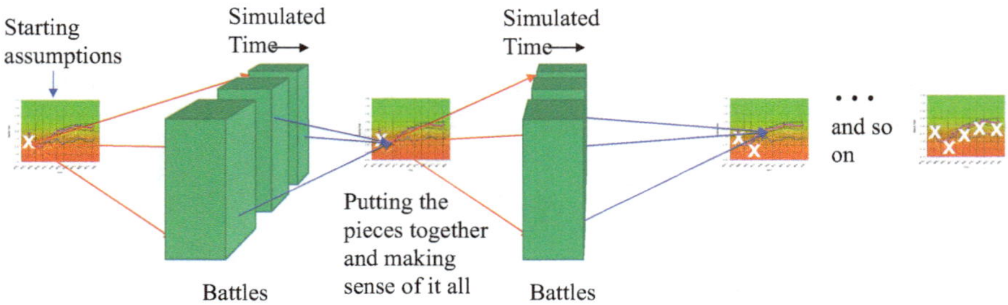

Fig. 3.9 Punctuated simulation

model with less improvement because of lessened economic activity, higher absenteeism, etc., and might restart the social model with somewhat less factionalism because of improved living conditions.

As shown in Fig. 3.9, the analysts run the second phase of the analysis, stop the simulations after the defined period, evaluate the results in the PMESII tool, and repeat the process as many times as necessary. The final result is a track of the output measures and the measure of success over simulated time that is based on the combined results of the tools, battles, and analysts' professional judgments. This entire process is repeated for each design point that is needed to complete the analysis.

Two sources of synergy are available in this conceptual methodology. The first is a synergy among the tools that is mediated by the PMESII tool and the analysts. The second is a synergy between the rigor of the tools and the professional judgments of the analysts, which allows for both compensation for inadequacies of the immature tools and rigorous grounding of the intuition and experience of the analysts.

Anthropological Modeling

The anthropological approach, like the psycho-social attributes construct, is bottom-up (Partnow & Hartley, 2010), rather than top-down as in the segmentation constructs. That is, the anthropologist starts with the people and looks for those areas that form culturally important categories for the people being studied.

- Leadership
  - What factors determine who becomes a leader?
  - What factors are involved in social schisms?
- Land use
  - What factors determine where a person/group settles?
  - When people move outside traditional territories, what parts of their lives remain unchanged?
  - What is the nature of the universe, as understood by the people?
  - Where do people fit into the cosmology?
  - What "beings" or forces affect reality, according to the cosmology?
  - What is the nature of the relationship between people and the natural universe?
  - How is individual and social health defined?
  - What causes physical or social illness?
  - How separate are the "natural" and "supernatural" realms, according to the cosmology?
  - What is the source of human power, according to the cosmology?
- Issues of social control
  - How do people deal with disagreements?
  - How does the group deal with those who break social norms (or laws)?

  - Who gets to decide punishment?
  - How and when is praise bestowed?
- Affiliation
  - How do the people define who is in the "in-group"?
  - Under what circumstances do people change affiliation?
  - What types of affiliation are strong? (E.g., familial, religious, linguistic, economic, friendship, marital)
  - What is the relative strength of the various types of affiliation a person practices during his/her lifetime?
- And so forth

An understanding of this baseline information as it relates to a social or cultural group is necessary if one is to understand what happens when things go wrong. For instance, what disruptions in the above dimensions are likely to result in violence? What internal mechanisms are in place to help the system return to a non-violent state? If none exist, then what would the effects be of externally imposed mechanisms? Understanding the various factors, the circumstances in which they are strengthened or weakened, and how they interact requires a multidimensional view.

Technical Modeling Approaches

There are many modeling approaches or techniques that can help with the problem of unconventional conflict. To avoid overwhelming the reader, the approaches are divided below into static-modeling, dynamic-modeling (but not time-modeling), time-modeling, other, and visualization and comprehension. Some of the approaches cross these divisions and many are commonly used together.

Static-Modeling Approaches

Static-modeling approaches or techniques may include time in their content, but the technique itself is independent of time. The entire model is a constant and does not change over time.

- **Conceptual Models**: A conceptual model is a composition of concepts that aids in understanding the subject of the model. It may, but does not have to, lead to a coded computer model.
- **Data Models**: Data models define the relationships among data types and, generally, provide metadata to support understanding the data.
- **Displays**: Displays may not initially seem to be models; however, they represent the results of modeling decisions because they display only parts of what might be known of a situation.

- **Flowcharts**: Flowcharts often represent conceptual models that will become coded models. However, some flowcharts constitute the final model. Some of these are process models that remind a decision maker of the steps he or she must take in making a decision. This can be extremely helpful in a DIME/PMESII environment where very little is simple and the complexities can be impossible to remember.
- **Lists, Taxonomies, and Ontologies**: Lists and their more complex relatives, taxonomies and ontologies, are useful for organizing what is known about a subject. They become increasingly valuable as the subject becomes more complicated. Their most obvious value lies in their use as memory aids. They have a more profound value as guides for creating more sophisticated models.
- **"PowerPoint" Models**: PowerPoint and other graphical design tools may be used to create the conceptual model that is the first step in creating a coded model. In some cases, the graphical model is sufficient for the purpose of allowing people to understand the relationships of selected items.

Figure 3.10 illustrates this latter use. It shows the complete corruption model developed in an NDU workshop on corruption in Afghanistan (Hartley, 2010b).

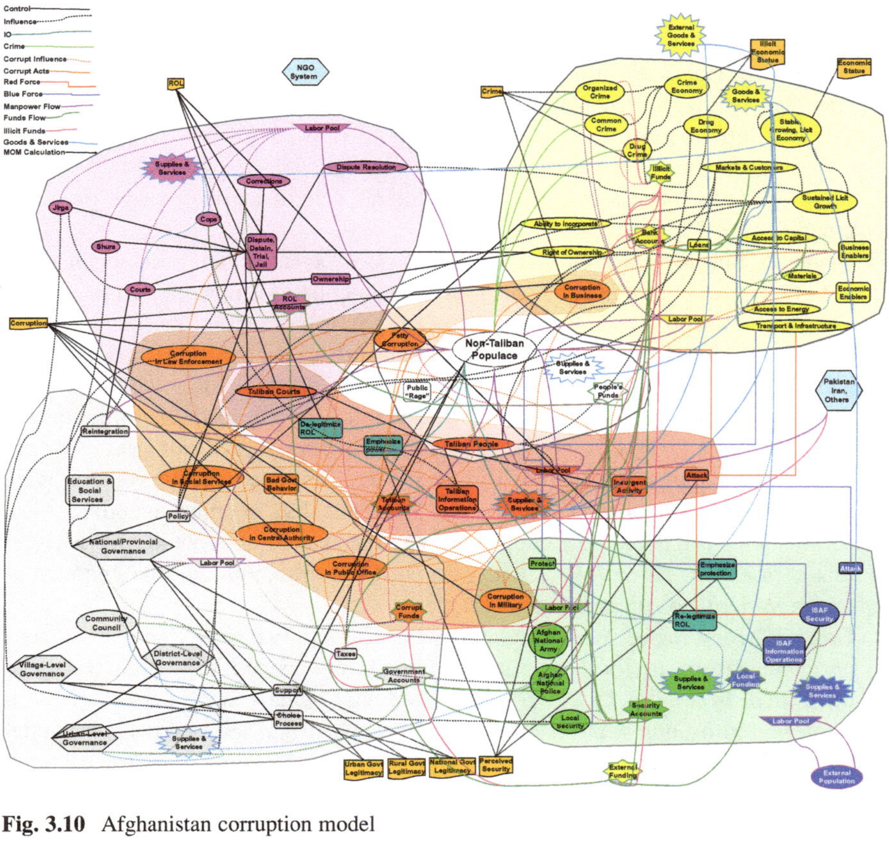

Fig. 3.10 Afghanistan corruption model

As illustrated, this is a very complex model; however, by using the build capability in PowerPoint, the model can be incrementally displayed, allowing for a more easily comprehended view of the model (and one in which the individual items can be read). This PowerPoint file can be downloaded from http://drdeanhartley.com/HartleyConsulting/NDU/NDUHSCB.htm. Page down to the diagram of the model and click on the download link below the diagram.

- **Process Models**: A process model describes the steps in a process. It can be linear or contain branches and loops. A flowchart is one type of process model.

Dynamic, But Not Time-Modeling Approaches

The dynamic, but not time-modeling approaches, have the capacity to change as they are run, and thus are not included as static approaches; however, they do not (generally) explicitly model the passage of time.

- **Algorithms**: Algorithms are usually components of larger models. Each algorithm computes a particular result from a set of inputs. However, some algorithms may be useful as stand-alone models.
- **Databases**: Databases are very often thought of as simple collections of data; however, relational databases can include much more than just a set of values. The relations among the values constitute models of a portion of the world. Complex databases can constitute quite sophisticated models of reality and, when dynamic processing code is included, can instantiate fairly powerful models.
- **Decision Support Systems**: Decision support systems are models that are designed to accept inputs and generate recommendations that support decisions. They may be implemented as databases or may use other coding practices.
- **Economic Modeling**: The economic models of Berman and Kuznar (Berman & Kuznar, 2010) used rational expectations to develop testable models.
- **Expert Systems**: Expert systems were created by the artificial intelligence community as a way of computing inferences from a number of simple propositions that capture the knowledge of experts. Initially, expert systems were written in the LISP computer language. Theoretically, an entire unconventional conflict situation could be modeled as an expert system. It is not known whether this is practical. However, certainly parts of the problem could be encoded as expert systems to support a larger model.
- **Game Theory Based Models**: Game theory-based models use the mathematics of game theory in their algorithms. The basic concept of game theory is illustrated in Fig. 3.11. Two sides, Red and Blue, have possible actions (represented by numbers on their respective axes) that they can choose, independently of the other. However, it is the pairing of the choices that determines the outcome. Some pairings result in outcomes that are very favorable to Red and some that are very favorable to Blue. The shape of the outcome surface in

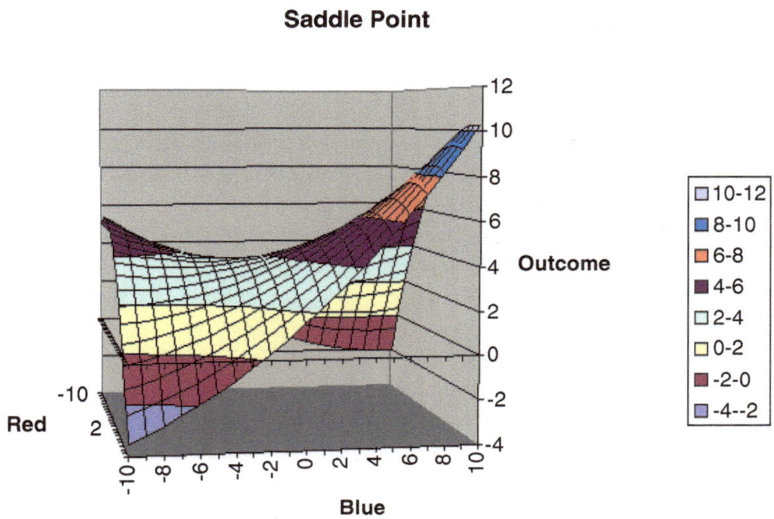

Fig. 3.11 Game theory saddle point

this case shows that a choice by either that can result in a very favorable outcome will also result in a very unfavorable outcome, depending on the other player's choice. Game theory says each side should play a minimax strategy that maximizes its own outcome while minimizing the other player's outcome. This choice results in the saddle point outcome in the middle of the surface.

This description covers the two-person zero-sum game. Because it has a linear payoff function, it can be solved by standard optimization methods, such as linear programming. Other games include multi-player games and games with non-linear payoff functions and require more complex optimization techniques for their solutions. Non-linear multi-player games, to the extent they can be formulated and validated, could be very useful in looking at unconventional conflicts with multiple players, not all of whom play by the same rules.

- **Influence Nets**: Influence nets are decision theoretic constructs. Wikipedia defines it as follows (Wikipedia, n.d.):

  "An **influence diagram (ID)** (also called a **relevance diagram**, **decision diagram** or a **decision network**) is a compact graphical and mathematical representation of a decision situation. It is a generalization of a Bayesian network, in which not only probabilistic inference problems but also decision-making problems (following maximum expected utility criterion) can be modeled and solved.

  "ID was first developed in mid-1970s within the decision analysis community with an intuitive semantic that is easy to understand. It is now adopted widely and becoming an alternative to decision tree which typically suffers from exponential growth in number of branches with each variable modeled. ID is directly applicable in team decision analysis, since it allows incomplete sharing of information among team members to be modeled and solved explicitly. Extensions of ID also find their use in game theory as an alternative representation of game tree."

An influence net is a model that computes a probabilistic result from a set of inputs. The concept can be reduced to a deterministic result simply by setting the probabilities to 1.0. In this case the value lies in the concept of multiple influences on an outcome or a decision. Influence nets can also serve as conceptual models for the construction of more complex models.

- **Network Analysis Models**: Network analysis identifies distinct elements or actors with the nodes (or vertices) of a network and the connections between the elements or actors with the links (or edges). It draws on theories and methods including graph theory from mathematics, statistical mechanics from physics, data mining and information visualization from computer science, inferential modeling from statistics, and social structure from sociology to draw conclusions about the model. Social network analysis concentrates on modeling social structures as networks. Dynamic network analysis adds the element of changes within a network over time (moving it into the time-modeling approach) (Carley, 2003).

  In 2010, David Siegel published a paper that provided a typology of qualitative network types to classify how different networks affect aggregate outcomes (Siegel, 2010).

  In 2013, Walker, Anglero and Davidson discussed recent developments in Network Discovery (Walker, Anglero, & Davidson, 2013). They said, "Network Discovery describes an empirical approach to assist in understanding several cause and effect relationships embedded within event and social networks." They described how they used their "research to address the important issues such as assessing the performance of key decision makers and discuss an approach to measuring the effectiveness of actions taken by decision makers through the use of these tools."

  In 2016, Neil Johnson et al. examined the online presence of supporters of ISIS and discovered an "ecology evolving on a daily time scale that drives online support" and provided a mathematical theory describing this evolution (Johnson et al., 2016). They also made a prediction "that development of large, potentially potent pro-ISIS aggregates can be thwarted by targeting smaller ones."

- **Optimization Models**: Optimization is the process of searching for the best value or searching for the input values that yield the best value. The best value may be a maximum value, a minimum value, or a constrained value, such as an integer value or a mini-max value in game theory. There are several techniques that may be used to find the best value, depending on the circumstances. These techniques include linear programming, nonlinear programming, integer programming, mixed-integer programming, and interior point methods. An optimization model will include such techniques to optimize one or more of its variables. Often optimization requires simplification of the real-world problem in order to create a solvable mathematical problem. This can reduce the validity of the corresponding real-world solution. Optimization may also suffer from a lack of robustness. That is, if the functional value of the inputs in a neighborhood of the optimal value differ greatly from the functional value at the optimum point, small variations in actually attaining the optimal value inputs in the real-world will yield results that differ greatly from the expected value.

- **Prediction Models**: Prediction models are simply models whose purpose is to predict something. As shown in Table 2.1, prediction models can use more than one technique to accomplish their purposes. Prediction models can be simple dynamic models or simulations.

  Wedgwood, Ruvinsky and Siedlecki mentioned the exploitation of information found in news data, developed for the DARPA Integrated Crisis Early Warning System (ICEWS) (Wedgwood, Ruvinsky, & Siedlecki, 2013). While this paper focused on other issues, they described a system that uses text-parsing, among other techniques, to extract data from electronic versions of news reports to support forecasting crisis situations.

  In 2013 Bell, Caglayan, Cassani, and Murdie described the Predictive Societal Indicators of Radicalism (PSIR) model (Bell, Caglayan, Cassani, & Murdie, 2013). "Using a regression model applied to a large number of drivers of conflict variables spanning numerous open source social science datasets, the PSIR model uses a novel negative residuals analysis technique" to forecast political violence.
- **Spreadsheets**: Spreadsheets can be simple calculating machines or with the addition of Visual Basic for Applications (VBA) coding can become complex models, some incorporating Monte Carlo simulations that generate sets of results based on distributions of possible results.
- **Text-Parsing Models**: Text-parsing models use rules to interpret text and convert it into a set of computable data. This is discussed under data creation in the "Gathering the Data" section of Chapter 7.

  In 2013, Cassani et al. discussed a methodology for automated semantic tagging of text data to support integration of socio-cultural knowledge (Cassani et al., 2013b). This methodology involved the creation of a semantic Wiki "that enables users to collect, organize, tag, search, browse, visualize, and share knowledge." It "provides intelligent and automated semantic tagging of field collected data imported to the knowledge portal generating new relationships within the data and making information more discoverable."
- **Tracking Models**: Tracking models include a time component as a data input, rather than a generated datum. At their core, tracking models are databases with time as a sequencing value. They may also include some kind of display to aid in interpretation. See the Interim Semi-static Stability Model in the Models section of this chapter and the paper by Goldstone (Goldstone, 2010).

Time-Modeling Approaches

Time-modeling approaches are simply the ones that explicitly model the passage of time and the changes to the model state that accompany the time change.

- **Agent-Based Modeling (ABM)**: The following discussion is taken from (Hartley, 2015a) and the discussion on V&V for Agent-Based Models in (Hartley, 2010c):
  Agent-Based Models (ABMs) are stochastic simulations that are built around dynamically interacting objects called agents. These agents are usually all of comparable size and scope. In some ABMs the agents are all identical copies of the same object, while in others there are two or more types of agents. The agents have preset attributes and behaviors; however, the choice of data inputs can dramatically change the model's purported real-world referent. For example, the same agents in an ABM can represent soldiers with a mission in one model and a set of mines in the ocean in another model. Generally, ABMs contain some spatial representation, (x, y) or (x, y, z) coordinates, that can be used to model real-world distances.

  The simulations that are most commonly referred to as ABMs (thin-agent models) have simple sets of rules; however, the simulations are dynamic. The rules mesh together during the simulation run and produce often surprising, emergent behaviors. The stochastic nature of the models and the sensitivity to small changes means that thousands to millions of runs are required to understand the range of behaviors, their frequencies, and associations with input data. The outcomes are non-linear, that is, not predictable beyond small time increments. Further, the results from N iterations will likely differ from the results of $N + M$ iterations.

  The emergence of complex behaviors from simple sets of rules is the principal reason for using ABMs. The claim is that just as the complexity of chemistry emerges unpredictably from the simple rules of atomic electron interactions, the best way to understand the complexity of human interactions is to investigate the emergence of complex interactions from simple simulation rules in an ABM.

  The ABMs of interest here are the models that include PMESII-type attributes, such as emotions, opinions, and social grouping valences. These attributes present problems because they are labeled as PMESII-type attributes; however, it is not clear that they actually represent what their labels claim. Because our understanding of the true relationships among such variables is poor, validating the code is difficult. Further, the claim is made by some ABM proponents that the primitives that make up the relationships are what should be modeled and that by observing the emergent behaviors, users can make correlations between these behaviors and real-world behaviors with these labels, producing the desired validation. Such models may also provide useful heuristics for real-world behaviors, regardless of whether the models can be validated or not.

  Some thin-agent models have a self-modification capability. This capability permits learning by the model, often using genetic algorithms. These ABMs support connected sequences of runs in which the results of previous runs are used to modify the behaviors in subsequent runs. These ABMs are adaptive and can generate the co-evolution of the behaviors of one or more groups of agents (sides).

The ideal thin-agent ABM would be a completely protean, content-free model in which any situation in any portion of reality could be modeled simply by changing the data and changing the human meanings attached to the objects and labels of the model. Thus the dots on the computer screen that the model uses to portray the agents can be thought of as individual soldiers or as floating mines. The rules governing movement toward or away from other agents can be thought of as social rules of liking or disliking or as representing the physical constraints of chains and wave action. This means that such an ABM is actually an environment for creating ABMs that model a particular situation. The protean ABM could be regarded as a meta-ABM for producing ABMs.

Actual ABMs are not ideal, having some categorical representation built in to provide a framework for the ABM construction. Thus a particular ABM might not be flexible enough to represent both soldiers and mines. Some ABMs might include more geographical representations than others. However, in general, most of the model is contained in the data and the human interpretation that is attached to the implementation. That is, most thin-agent ABMs are actually environments.

The use of the simpler agent-based models introduces an analysis problem. First, the generation of thousands or millions of model runs yields extremely large amounts of data that must be analyzed. Second, the number of variables to be examined, while typically smaller than in complex models (perhaps on the order of 20–30 variables versus hundreds to a thousand variables), still represents an impossible task if a full analysis is desired. A full factorial design for five levels of 25 variables (to estimate non-liner effects) yields approximately 3×10^{17} runs. Even if the ABM only takes 1 s to run, it would take one computer about 9.5 billion years to complete all of the runs. Using a supercomputer with a million processors reduces the time required to 9500 years. The Naval Postgraduate School (NPS) Simulation Experiments & Efficient Designs (SEED) Center has developed a much more efficient methodology for design of experiments that yields the equivalent information in 2 or 3 hours (NPS SEED) (Cioppa & Lucas, 2007).

The more complex ABMs (thick-agent models) incorporate variable behaviors and have complex rules. Or they may have no explicit rules at all, instead having large amounts of code that substitute for rules.

- **Games and Wargames**: In modeling parlance, games or wargames are human-in-the-loop (HITL) models, generally simulations (see below). That is, the model relies on human beings to supply a significant part of the logic, usually the decision-making part. The rest of the game consists of adjudication and support processes, such as information display. Games can be as simple as a board game, with a spinner or die performing random adjudication, or as complex as a multi-model computer system, with computer algorithm adjudication. In the latter case, the humans playing the game may be directly using the computers or may be in the field, relaying orders to operators who interface with the computers.

In 2015, Mike Ottenberg discussed issues relating to games and wargames (Ottenberg, 2010). A major problem with games is the difficulty of the verification and validation (V&V) process. Another problem involves the composition of the threat forces: using a fictional opposition, another identical force or a mirror image force, all create difficulties in estimating the value of the results. The lack of moral hazard in making decisions or exhibiting behavior brings into question the validity of the overall model. And the time and resources required to set up a wargame for operational analysis mean that usually only a single run of a given scenario is made, leaving open the question of what differences in results might have been seen with a different run.

- **Forecasting or Predicting Models**: As mentioned above, these models can also be simulations. See the paper by Goldstone for differences with tracking models and problems in developing forecasting models (Goldstone, 2010).
- **Simulations**: Simulations are models that explicitly model the passage of time and the changes that accompany that passage. Simulations may be deterministic (always produces the same output, although chaotic regimes are possible) or stochastic (the simulation uses some form of a pseudo-random number generator to produce one output from a distribution of possible outputs, with different results in different runs, also referred to as Monte Carlo simulation). Simulation time may be time-stepped (the simulation clock is incremented by a fixed value to simulate the flow of time, although portions of the simulation may use a smaller time-step value to achieve finer-grained results) or discrete-event controlled (events are placed in an event queue, each with a scheduled time of occurrence; events are pulled from the queue based on earliest scheduled time; the simulation clock is advanced to that time and processing continues). System Dynamics is a particular kind of time-stepped simulation that is based on simulating levels of variables and flows among them. Typically, the first step is to draw a diagram, such as Fig. 3.16, in the FM 3-24 Counterinsurgency Model in the next section. The diagram is converted into differential equations that are evaluated as difference equations with a fixed time-step.

  Most analysts think "Monte Carlo simulation" when they see the term "simulation," but the U.S. Defense Modeling and Simulation Office (DMSO) used "simulation" to apply to anything except the real thing, and subdivided it into live (real people operating real systems firing real munitions), virtual (real people operating simulated equipment), and constructive simulations. It is a constructive simulation that we think of as a "model." There are also hybrids, namely combinations of live and constructive simulations; "simulation" in support of exercises is one of these. In that case, its intent is verisimilitude (the appearance of reality) rather than any form of physical reality.

  Another issue with the use of the term "simulation" arises with consideration of the understanding of the "results" of a simulation. A Monte Carlo simulation generates multiple instances (results) as its default mode and (generally) includes the display and analysis software to make sense of the results. Many other "simulations" with stochastic elements are really stochastic instance generators; any attempts to use them to analyze multiple instances rely on analysis that lies completely outside the "simulation".

In 2013 Neil Johnson described a methodology for developing a minimal "mechanistic" model of conflict from spatiotemporal datasets and current narratives (to create stylized facts) (Johnson, 2013).

- **Virtual Environments**: Virtual environments are simulations with HITL. However, while games may model intervals between each human intervention, virtual environments model the continual passage of time while the human is connected and concentrate on providing an immersive environment so that the human experiences the model's changing situation as if it were reality.

Other Approaches

"Other approaches" is the catch-all category.

- **Frameworks:** Frameworks are environments that "know" about a particular domain and provide support for building models in that domain. A proposal to use the IW Ontology (described later), together with a user interface would create such a framework for unconventional conflicts (Hartley, 2016a).

  In 2013 Pynadath and Marsella described the PsychSim agent-based social simulation framework (Pynadath & Marsella, 2013). "PsychSim takes a decision-theoretic approach to modeling Theory of Mind, giving its agents a rich space of beliefs and preferences. PsychSim also uses a unique piecewise linear representation of language that allows it to reason backward from observed behavior to infer consistent parameter settings."
- **Modeling Environments**: Modeling environments can range from code-writing environments, which provide the cues and language checking functions to support creating well-formed programs, to domain-specific environments (frameworks) that "know" about a particular domain and provide support for building models in that domain.
- **Multi-Modeling**: Multi-modeling consists of using multiple, inter-related models on the same problem. "The critical issue in multi-modeling is ensuring that the scale and boundary conditions for those models that are locked are consistent (Carley, Morgan, Lanham, & Pfeffer, 2013)." They described four forms of inter-relation (docking, collaboration, interoperability, and integration) and said that multi-modeling "supports in-depth analysis by enabling cross-validation of results through triangulation, insight into implications at multiple levels of granularity through cross-model consistency, and model advancement through re-use across multiple domains."
- **Systems of Models**: Systems of models are collections of models. They may be closely connected, that is with automatic passing of data from one model to another, or loosely connected, requiring human intervention to pass data among the models. Multi-modeling is a special case of a system of models with replication

of function in different models. A system of models may be a single large model designed as a system to simplify the creation process; however, systems are often the result of connecting models built separately for other purposes. In any case, the connections between models in the system are critical failure points.

An Oak Ridge National Laboratory project on an Army system of systems project identified levels of connection (Hartley, Loebl, Rigdon, van Leeuwen, & Harrigan, 2004). "**Linking** means that the output from one function [read as "model"] can serve as input to another function syntactically. Linking is necessary, but not sufficient, to specify which functions really are interoperable. Two functions are said to link or form a linkage or are said to be capable of linking if the output of one is of the correct form to be used as input to the other." "**Interoperability** means that the output from one function can serve as input to another function, meeting the semantic needs, as well as the syntactic needs, for input to the second function." Normally two independent models will not be capable of linking, much less interoperating. "An **interface** is a function [read "small model"] that resolves incompatibility between two functions. For an interface, the completeness of the resolution is a concern." The report identified a situation in which interoperability is not sufficient for full system validity even if all of the models were fully valid. In that case, the connections were correct syntactically and semantically; however, the internal granularity of the representations among three connected models was not identical. This meant that no interface change could correct the problem. The only solution was to change the data for the finest-granularity model so that it replicated the coarser granularity of the other models.

Visualization and Comprehension Approaches

Visualization and comprehension techniques support other modeling approaches in the effort to increase the models' value by making them more comprehensible to users. It should be noted that using visualization to support comprehension applies to more than just modeling. Edward Tufte produced a book on visualizing quantitative information, which is the classical work on statistical graphics (Tufte, 2001). As an example of a great graphic, it includes the amazing graph of Napoleon's invasion of Russia that shows the diminution of Napoleon's force size as a function of the distance from its origin to Moscow—and back.

Jonker and Wright discussed visualization and comprehension (Jonker & Wright, 2010). They began with, "(t)he Achilles' heel of societal models, nearly universally, is their inability to convey their computational results to the human user." Perhaps the most basic purpose of a model is to increase the user's comprehension of something.

The point of their article was that visualization techniques are a means to convert the un-comprehended into the comprehended. They divided the roles of visualization into two categories, interactive data or information visualization and visual analytics, which joins "interactive visualization with analytic reasoning and computational methods in order to derive and convey insight from massive, dynamic, ambiguous, and often conflicting data." Their article explored a number of visualization techniques in several model settings.

Wedgwood, Ruvinsky and Siedlecki mentioned the exploitation of information found in news data, developed for the DARPA Integrated Crisis Early Warning System (ICEWS) (Wedgwood et al., 2013). This paper focused on the problem of providing a robust transparency framework. They "provide an example of how the framework can be used to support traceability of HSCB model artifacts, increase user confidence in HSCB models and support understanding of the model dynamics . . . ."

Models

I know of more than 150 models (using a loose definition of the term) that might be useful in the unconventional conflict domain. I selected the individual models and modeling systems below because they display the range of unconventional conflict models and some of the issues they present and because I have had some experience with them. Some have been or will be retired and others are still being used. That status is immaterial here. There is something to be learned from each one.

Joint Theater Level Simulation (JTLS®)

JTLS started as a combat model to support theater-level operations planning. Because of its interactive construction (human-in-the-loop), its evolution to support training was not difficult. The addition of the capability to address HA/DR operations in 2013 has led to the inclusion of HA/DR scenarios in its use by Taiwan in Poland (Roland, 2016). The following description of JTLS is taken from the JTLS website (Rolands & Associates Corporation, 2016).

> The Joint Theater Level Simulation (JTLS®) is an interactive, Internet-enabled simulation that models multi-sided air, ground, and naval civil-military operations with logistical, Special Operation Force (SOF), and intelligence support. JTLS development began in 1983 as a project funded by the U.S. Readiness Command, the U.S. Army Concepts Analysis Agency, and the U.S. Army War College. The simulation was originally designed as a tool for development and analysis of joint as well as combined (coalition) operations plans. Today, JTLS is always used as a training support model that is theater-independent and does not require a knowledge of programming to operate effectively.
>
> The primary focus of the JTLS system is conventional joint and combined operations at the Operational Level of War as defined by the Joint Staff's Universal Joint Task List. JTLS explicitly models air, land, sea, amphibious, and SOF operations. The simulation supports limited nuclear and chemical effects, low-intensity conflict, pre-conflict operations, as well as support of Humanitarian Assistance and Disaster Relief (HA/DR) operations.

Unit Order of Battle (UOB) Data Access Tool

The Unit Order of Battle (UOB) Data Access Tool (DAT) was designed to provide simulation developers with consistent and authoritative order-of-battle information. It tailors forces and equipment to specific missions across the levels of war. The intended purpose is to automate the initial data load process and consists of three main components: a Data Interchange Format (DIF) file, a data extraction tool (DET), and a set of authoritative data sources (ADS).

The DIF presents order of battle information from all sources in a single understandable format, eliminating the need for simulation developers to understand the minor differences between Service-specific and foreign order of battle data sets. (An order of battle data set contains the description of a military force with its subordination structure and the personnel and equipment that are part of the force.) The format is based on data elements that have been formally reviewed and adopted by the DoD Data Dictionary System. Data instances are enumerated using a rigorous data coding specification. U.S. UOB is enumerated using a Unit Identification Code, a National Stock Number, and/or a Service-specific occupation code. Non-U.S. UOB is enumerated using appropriate Defense Intelligence Agency coding specifications.

The DET provides unit UOB data in the interchange format. This tool features a graphical interface that allows users to browse UOB data and select individual units to form a scenario task force to start a simulation-supported exercise. Other information covers unit equipment, aircraft resources, and personnel. The tool supports organizing the reporting hierarchy of a task force, including adding specific or generic units. Further, the tool supports "rolling up" subordinate units into a parent unit, which is important for simulations that operate at aggregation levels above the basic unit (the smallest unit in the database), and supports equipment and personnel "mapping." The tool allows for local data editing and scenarios can be saved for later use.

ADS include the Conventional Forces Database that covers order-of-battle information on U.S. units (classified SECRET) and the Military Intelligence Data Base that covers order of battle information on foreign units (classified SECRET). A Generic Data Base covering unclassified U.S. and foreign forces is also provided. For OOTW use the UOB DAT requires inclusion of sources for allied forces, International Organizations (IOs), and NGO/PVOs.

Lessons Learned Information Warehouse (LLIW) and Mine Awareness

The Lessons Learned Information Warehouse (LLIW) is a Canadian product (Champoux, Trudel, & Thibault, 2004). It contains a tremendous amount of information concerning OOTW operations and a powerful search engine, including:

- Historical operations by name;
- Training standards, reports, after action reviews, task descriptions;
- Lessons learned from U.S., British, and other allies, and *Military Review* articles;
- Terrain and intelligence analyses, such as the Central Intelligence Agency (CIA) World Factbook;
- Publications and doctrine concerning OOTWs; and
- Search and annotation support for LLIW contents.

Produced by the National Defense Mine/countermine Information Center, the Canadian Forces Landmine Database (CFLD) has been designed as an information tool for those individuals and agencies engaged in Landmine Clearance and Mine Awareness Training, or by military forces in preparation for deployment. It is a tool for discovering the types of mines found in an Area of Responsibility (AOR) or identifying a mine by physical characteristics and displaying information about the mine. The mine database is UNCLASSIFIED and has been compiled from numerous agencies and individuals from around the world to ensure that the information provided is the best available. It has pictorial representations of over 350 mines, cross referenced for tactical employment. This tool is also included in the LLIW (Fig. 3.12).

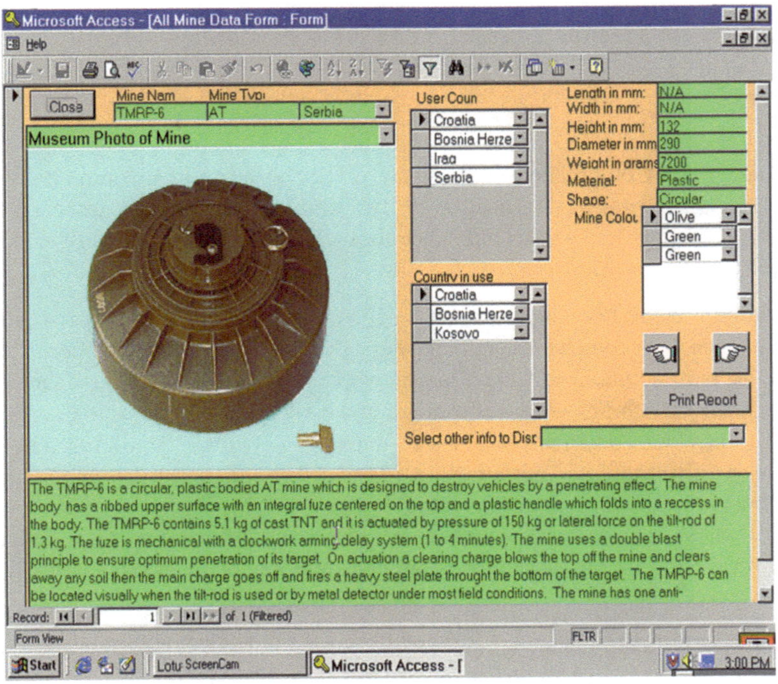

Fig. 3.12 CFLD mine awareness tool

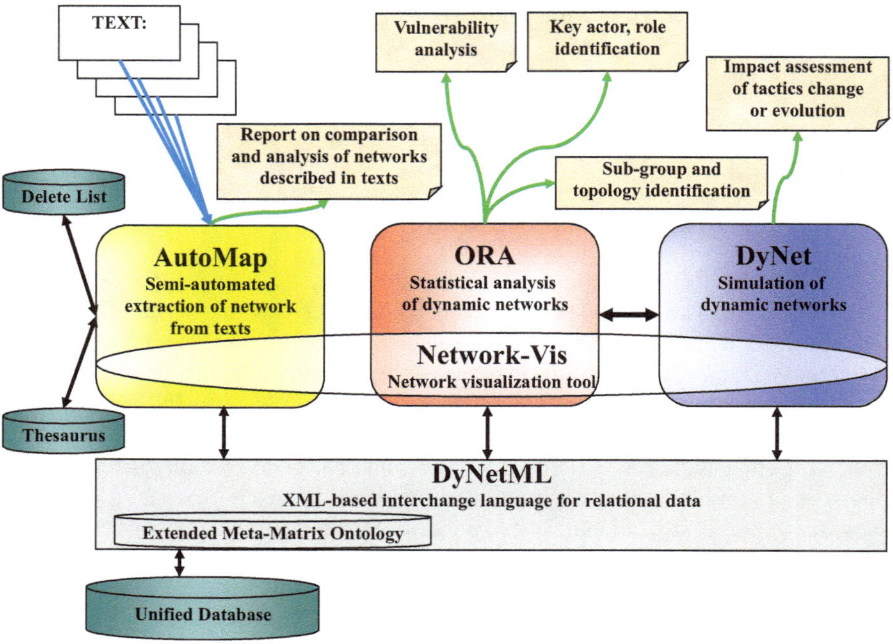

Fig. 3.13 Capability and interoperability of CASOS tools

CASOS

The Center for Comprehensive Analysis of Social and Organizational Systems (CASOS) is an organization headed by Dr. Kathleen Carley at Carnegie Mellon University. Figure 3.13 shows the major tools and their interactions in the CASOS system (Carley, 2004).

- AutoMap semi-automatically extracts network information from text and provides reports, such as comparison and analysis of the networks described in the texts;
- DyNet is a simulation of dynamic networks that allows the assessment of the impact of such things as tactics changes and evolutionary changes to the network;
- ORA provides statistical analyses of dynamic networks, supplying reports on vulnerability analyses, identification of key actors and their roles, and sub-group and network topology identification;
- Network-Vis is a network visualization support tool for the other tools; and
- DyNetML is an XML-based language that supports the interchange of relational data.

Figure 3.14 illustrates the type of output available from ORA, here showing an Al Qaeda network. Each node represents a person. The actual ORA output shows the person's name beside the node; however, the names have been removed for clarity (except for two prominent actors). The links show the known connections among the nodes.

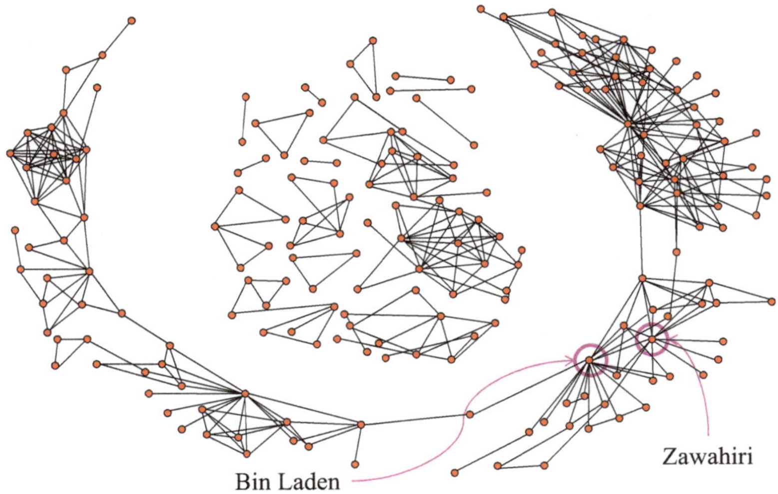

Fig. 3.14 Simplified ORA output

Interim Semi-static Stability Model (ISSM)

The ISSM describes the impact on the measure of political effectiveness (MoPE), "civil stability and durable peace," of the values of a set of factors within a single country. This is an instantaneous impact and does not allow for any feedback loops or actions over time (Hartley, 2006a).

The ISSM is implemented as a Microsoft Excel spreadsheet model. The current version is static, meaning that once an input has been entered, the change to the output is calculated and will not change over time. This can provide a useful approximation for some applications, as described below; however, it is only an approximation. In reality, when a change is introduced, there is a characteristic lag before any effect is seen and a time interval during which the effect grows to its maximum. These dynamic attributes are not modeled, hence the "static" in the model name.

The connecting logic and mathematics, while plausible and derived from reasonable sources, have not been validated. Hence the "Interim" in the model name. The structure is sufficiently flexible to permit ease of modification of the mathematical model as the situation warrants.

The ISSM is documented in a User's Guide (Hartley, 2006b) and an Analyst's Guide (Hartley, 2006c). This model was inspired by the model described in Doing Windows: Non-Traditional Military Responses to Complex Emergencies, Bradd C. Hayes and Jeffrey I. Sands (Hayes & Sands, 1997). However, nodes have been added; some definitions have been changed; and some connections have been changed. In addition, the connecting logic, while similar, is not identical.

The simplicity of the ISSM supports use at several echelons of military and civilian activities. In its most straightforward use, the ISSM gives an estimate of the ground situation in terms of a top-level Measure of Political Effectiveness (MoPE), namely the status of the assertion that civil stability and durable peace exist. Thirty-four user inputs are required. The values of these inputs can be obtained by intuition, based on observations of the current situation, or they might be obtained through sophisticated data collection by teams in the field. Naturally, the believability of the output value will be influenced by the believability of the input values. In any case, the calculation of the output value, given the input values, requires only minutes for input and less than a second for the spreadsheet calculation. The ISSM has also been used as an interpreter of the results of a simulated situation for a study. In this mode of operation, the simulation provides the surrogate reality and the ISSM measures and tracks the strategic results.

In both modes of use, the ISSM provides a preprocessor to support custom logic connecting available data with the inputs required by the ISSM Main workbook. The ISSM also provides a postprocessor to support custom logic connecting the available outputs with customer-defined Measures of Merit (MOMs). For those interested in more details, the pdf file of the ISSM v4.00 Users' Guide can be downloaded by using the link to the first reference cited at the bottom of the webpage http://drdeanhartley.com/HartleyConsulting/TOOLBOX/issmdesc.htm.

Pythagoras

Pythagoras is an agent-based simulation, developed by Northrop-Grumman (Bitinas, 2009). Like many other agent-based simulations, Pythagoras is an extremely fast-running, very simple, but protean stochastic simulation. The agents can sense their environment, direct other agents, communicate with other agents, move, and influence the environment. Its protean nature means that it can be configured to represent an almost unlimited number of situations and is not limited to military domains. Its speed supports making thousands of runs to examine vast portions of its parameter space and develop statistical evidence on results.

Diplomatic and Military Operations in a Non-warfighting Domain (DIAMOND): US

The United Kingdom's (UK) Ministry of Defence (MOD) Science and Technology Laboratory (DSTL) developed DIAMOND. It was designed to model joint OOTW campaigns. It models the treatment of non-combatants and their behavior, infrastructure impact (damage, repair, reconstruction), and multiple interactions between factions. Early use of the DIAMOND model was based on UK operational experience. The DMSO team has modified the model to use U.S. forces and equipment, made the input screens more user-friendly, and provided reports tailored to U.S. warfighter needs.

DIAMOND is a moderately fast-running, high-level, partially stochastic, object-oriented simulation of peacekeeping, peace enforcement and humanitarian aid operations (collectively called Peace Support Operations (PSO)). The major aspects of its technical design include:

- A simple node and arc network that provides a graphical representation of the region and environment, allowing the model to represent key areas of interest. Key facilities, such as airports and civilian shelter, can be represented.
- The representation of key actors and contributors to PSO by use of Entities. These represent the capabilities and behaviors of military units, civilians, non-military organizations, and the leaders or commanders for each. Entities interact with each other and the environment and exchange or consume key commodities such as food, fuel, and ammunition.
- A mechanism to organize Entities into common 'parties' that represent specific organizations or common groups within a scenario. These parties have an appropriate command structure and communications network to facilitate the allocation of missions and flow of intelligence throughout the party. Parties have relationships with one another that define their interactions.
- A mechanism to represent each party's concept of operations by nesting objectives in a series of plans and for those objectives to consist of a series of missions that entities can prosecute during a campaign. Commanders within a party allocate resources to achieve their objectives in line with the sequence of plans and the simulation terminates when a set number of parties achieve their end-state conditions or when a predetermined period of time has elapsed.
- A mechanism (referred to as negotiation) to obtain access to an area denied to one party by another and to allow multi-party co-operation to achieve aims and objectives without having to rely entirely on their own resources.
- A mechanism to allow entities to gain information on their environment and other entities through sensing, interactions and communication during model run. This information is organized into a local picture that allows those entities to make informed decisions on how they should prosecute their missions and activities delegated to them by their superior commanders.

Joint Conflict and Tactical Simulation (JCATS)

JCATS is an interactive, high-resolution, entity-level conflict simulation that models joint (air, ground, and sea), multi-sided combat on high/low-resolution digitized polygonal terrain. The Lawrence Livermore National Laboratory developed JCATS with the U.S. Joint Warfighting Center as the simulation proponent. An object-oriented simulation, JCATS supports a server/client environment that is designed to normally run in a distributed environment; however, a single computer can serve as both a server and a client. JCATS can model joint integrated capabilities, providing a model capable of providing joint and combined entity-level play.

JCATS is used predominantly by tactical battle simulation centers, military schoolhouses, military experimentation labs, and security forces for brigade-level exercises (training), analysis, and planning/rehearsals. However, with its aggregation capability and ability to support multi-story building plans in an Urban Warfare Environment, JCATS has attracted the attention of agencies conducting experimentation, planning, and rehearsals in the urban environment.

Senturion Model

The Senturion model simulates shifts in opinion over time (Abdollahian, Baranick, Efird, & Kugler, 2006; Abdollahian, Kugler, Nicholson, & Oh, 2010; Efird, 2009). The Senturion methodology and software address highly complex, politically driven issues characterized by multiple stakeholders attempting to influence each other and outcomes.

Figure 3.15 (used by permission from Dr. Mark Abdollahian) shows how the Senturion model iterates to reach a convergence of opinions. In the illustrated scenario, each party is represented by a colored line, with the vertical axis representing the opinion value, and all parties converge to a single opinion. In other scenarios, some of the parties converge to one opinion, while others converge to other opinions.

Fig. 3.15 Senturion output

FactionSim Model and PMFServ

The University of Pennsylvania FactionSim model is an agent-based model of the social and organizational roles that exist in an area of operations. In one instance, it uses the proprietary Performance Moderator Function Server (PMFserv) agents to model a terrorist group, a leader council, and four population groups (Silverman, 2012). Silverman gives a tutorial on how it works in an earlier paper (Silverman, 2010).

FM 3-24 Counterinsurgency Model

The U.S. Army's field manual on counterinsurgency, FM 3-24 (HQ DA, 2006a), developed an extensive analysis of the important factors in insurgencies and the factors in countering them. Following the 2005 development of FM 3-24, a J8 team of the Joint Staff created a systems dynamics model of these factors and their relationships (Pierson, Barge, & Crane, 2010). The system dynamics diagram is shown in Fig. 3.16.

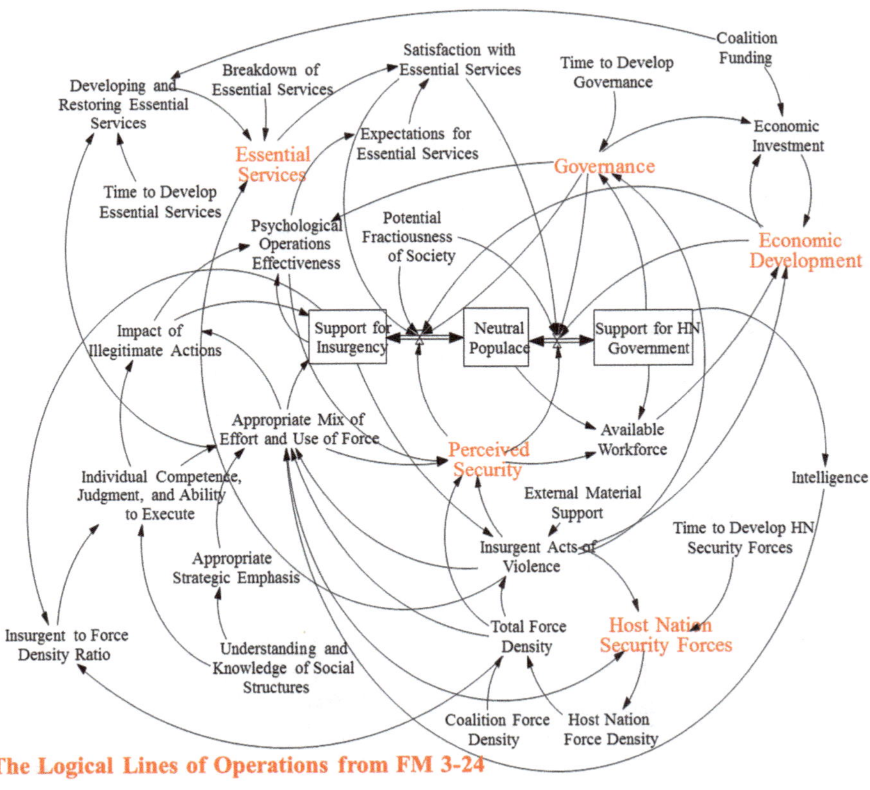

Fig. 3.16 System dynamics model of COIN operations

Synthetic Environments for Analysis and Simulation (SEAS) Model

The Synthetic Environments for Analysis and Simulation (SEAS) is a large (millions of agents) PMESII simulation of societies created by Simulex, Inc. SEAS models human behavior by combining concepts of individual self-efficacy with the social context that leads to social normative beliefs and with the cultural environment that leads to attitudes. The figures and table are taken from a 2009 presentation at the National Defense University and are used by permission of Alok Chaturvedi (Chaturvedi, 2009).

As shown in Fig. 3.17, individuals are represented and combine into organizations and institutions. Along with these, the geography and the infrastructure provide the support for representing the PMESII systems. SEAS does not model kinetic events. However, it does recognize the impact of kinetic events on the PMESII systems.

Figure 3.18 illustrates the cognitive sophistication of the SEAS agents. Each individual has its own cognition, based on cognitive psychology, neuroscience, and artificial intelligence. Each individual also has its own sensors, based on communication theory, and traits and actions, both based on psychology. Organizations and institutions also have cognition, sensors, traits, and actions; however, these agents also have members and leaders.

Infrastructure agents are also represented with their own version of traits, sensors, goals, and actions. They carry out production, trade, and distribution functions.

Figure 3.19 shows how these agents connect with the rest of the simulated world, performing various roles and using senses and actions to interact. The traits are represented as DNA, with the "DNA sharing" indicating how organizations and members are related.

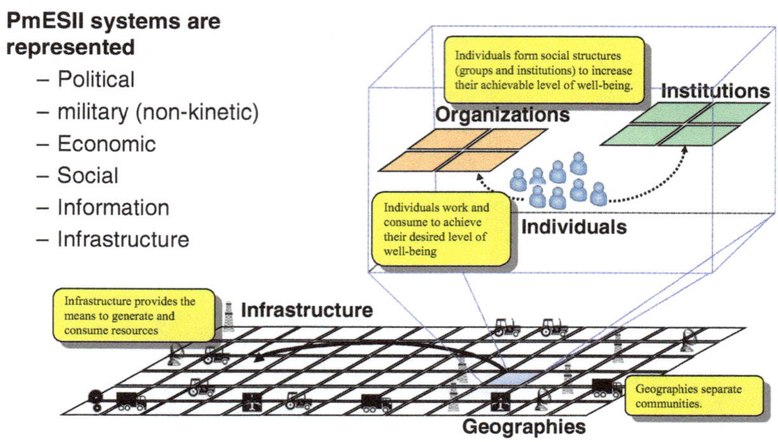

Fig. 3.17 SEAS representation of societies

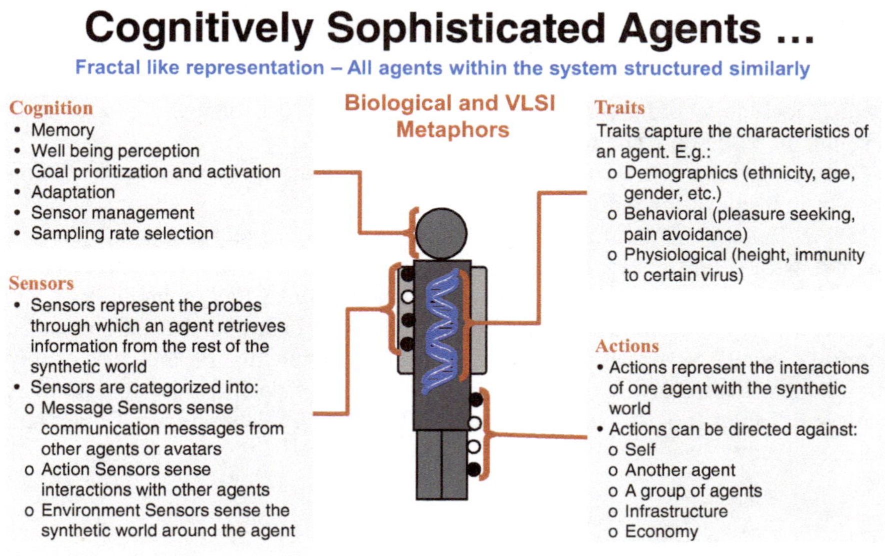

Fig. 3.18 SEAS cognitively sophisticated agents

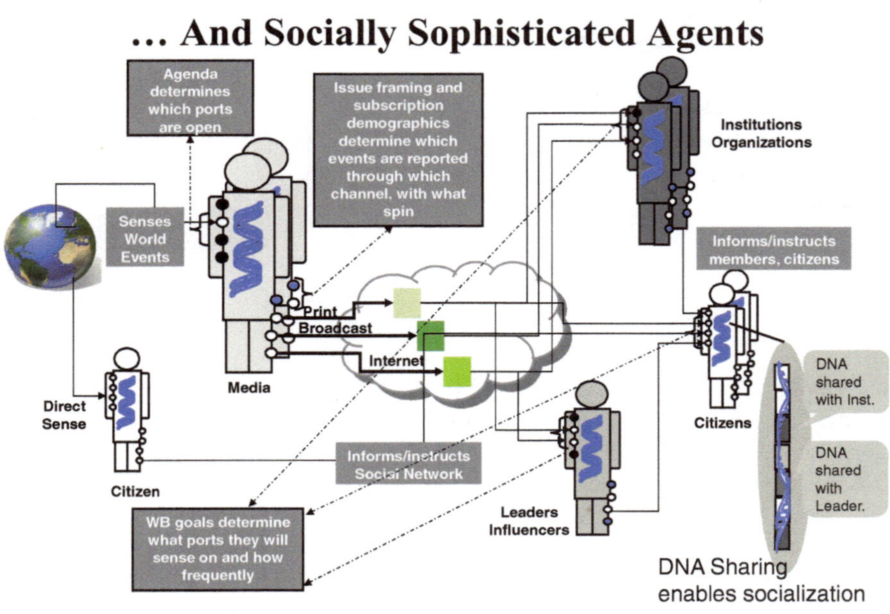

Fig. 3.19 SEAS socially sophisticated agents

This "social sophistication" is based on several human behavior theories. For example, one of the principal sets of theories is subjective well-being, which is used in economics, social psychology, political science, and political psychology. Other theories include set of happiness, preferences on goals, decision making, references (evaluation based on some reference), and mood states.

Table 3.1 differentiates the individual, organization, and institution modeling content in terms of the individual + PMESII categories.

SEAS Reference World (RW) uses semantic mining of real-world information sources with sophisticated text-parsing to create data on activities, entities, and metrics for its simulation. Figure 3.20 illustrates the SEAS RW technology.

The SEAS model requires extensive quantities of data and that data must be carefully vetted and cross-referenced. Thus the complexity of this figure only suggests the nature of the data activities. A sample of data sources for SEAS is shown in Tables 7.6, 7.7 and 7.8, found in Chapter 7, Gathering the Data, Data Sources, below.

Peace Support Operations Model (PSOM)

The Peace Support Operations Model (PSOM) was created by DSTL in 2006. It is a time-stepped simulation that is partially agent-based. It is properly a wargame, as it employs HITL. It is mostly deterministic. Because it is very fast-running, it can be run thousands of times using SEED design of experiments (holding human decisions constant for each set of runs) for distillation.

Table 3.2 shows the correspondence of PSOM outputs and PMESII indicators. The highlighted outputs are ones the PSOM user manual says are primary MOEs (Marlin, 2009).

DEXES II System

The Deployable Exercise Simulation (DEXES II) is a computer system developed by the University of Colorado Denver under contract to Northrop Grumman Corporation (NGC) for use by the United States Southern Command (USSOUTHCOM). It is derived from the earlier DEXES model, described by Woodcock (Woodcock, 1996). The purpose of DEXES II is to support analysis and exercises (training) in the socio-political domain. Topics of interest within this domain include armed conflict, civil unrest, criminal activities, epidemics, population displacement, and public opinion on various topics. The Principal Investigator for the model creation is Dr. Loren Cobb of the University of Colorado.

The DEXES II system is divided into three units. The simulation portion consists of the Deployable Exercise Simulation (DEXES II) model, which is derived, in part, from the earlier DEXES model. Control of the system is through a webpage,

Table 3.1 SEAS modeling

	Individual	**Organization**	**Institution**
Individual	Civilian Behavior Model (Culture-specific Normal, Transgressive, altruistic), Wellbeing	Family, Friends, Informal Groups	International Relations Model
Political	Citizen Organization Subscription Model, Political Leader	Political Organization, Recruiting, Training, and Retention	Political Parties, CX Government, COI Government, International Organization -- United Nations,
Military	Insurgency Model, Global Terrorism Progression Model, Terrorist Leader Model, Combatant Model, Will to Fight (WTF)	Terrorist Organization, WTF, Militia, Paramilitary, Rebels, Secessionist, Muti-national	Military Institution Model (Air, Land, Sea); WTF Model
Economy	Leaders, Influencers, Labor, Consumption Model, Production Model, IO models	Economic Sector, Business Organizations, Multinational Corporations, Informal Economy	Global Economy Model, World Bank, IMF, Trade Regimes (WTO, NAFTA)
Society	Social Group Model, Citizen Organization Subscription Model, Citizen Unrest Model Epidemiological Model, Religious Leader Model, Social Leader Model	Religious Organization, Social Network, Charitable Organizations, Humanitarian Organizations, Anti-social (Organized crime, syndicates)	Social & Religious institutions - (churches, mosques, temples), Non-governmental organizations (UN, Red Cross), Education
Information	Citizen Perception Model, Rumor Propagation Model, Media Subscription Model	Print Media, Broadcast Media, Online Media (Blogs)	Information and Broadcasting, Entertainment
Infrastructure	Individual Infrastructure, Infrastructure Dependence	Multinational Corporations	Infrastructure Management

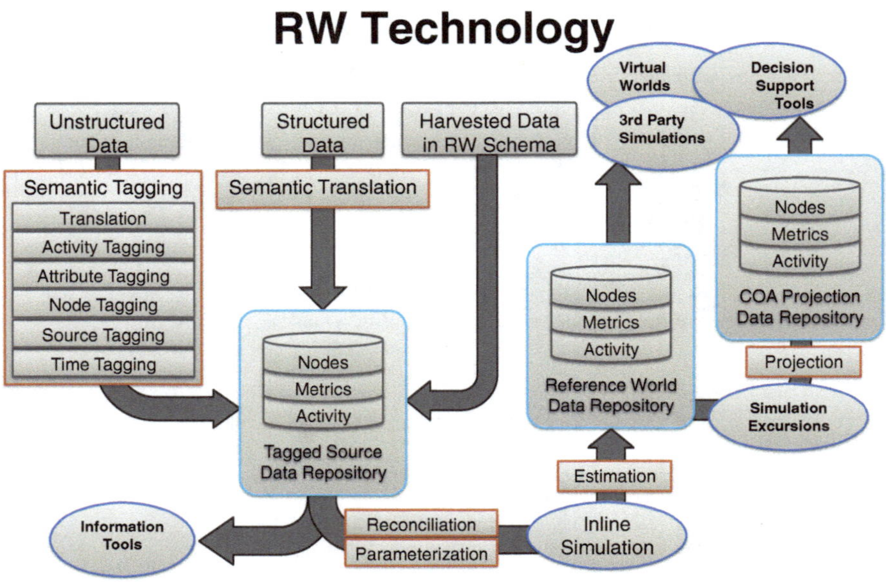

Fig. 3.20 SEAS reference world

Table 3.2 PSOM outputs and PMESII indicators

Indicator	Corresponding PSOM output
Political legitimacy	Population consent toward own Government
	Security
	Rule of law
	Corruption
	Provision of essential services
Military	Casualties
	Contacts
Economic (state functionality)	Production
	Reconstruction
	Income
	Human capital
	Growth rate
Social	Fear
	Rule of law
	Economic factors
Information	Human Intelligence (HUMINT)
	Headlines
Infrastructure	Infrastructure
	Economic production
	Human capital

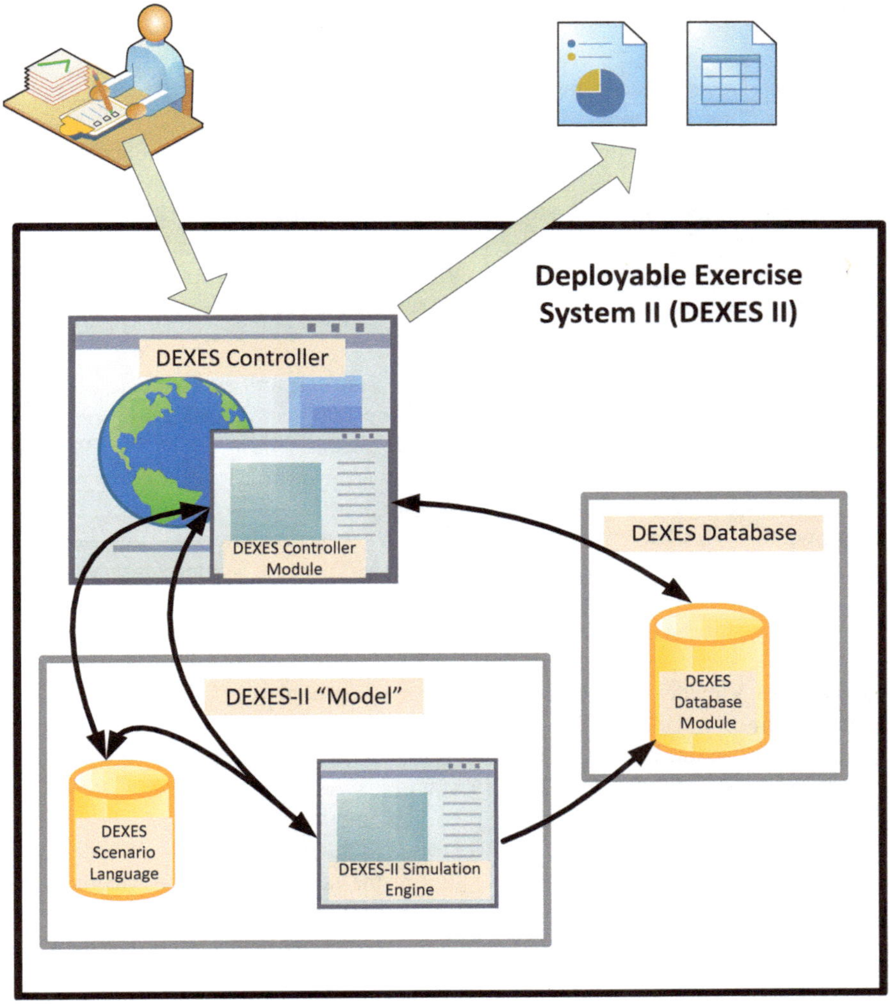

Fig. 3.21 The DEXES II system

the DEXES Controller, which supports user input and documented output. The simulation output is maintained in the DEXES database. The DEXES II model contains the DEXES II simulation engine and the DEXES Scenario Language. Figure 3.21 illustrates the make-up of DEXES II and its internal and external connections (Hartley, 2012).

Validation work on DEXES II is also covered in the discussion of VV&A in Chapter 6.

OOTW Toolbox Preliminary Design

In 2001, the Defense Modeling and Simulation Office (DMSO) began a project to create a set of OOTW tools.

The first step was to define the possible toolset (Hartley, 2001c). The possible tools were divided into short-term, mid-term, and long-term time frames for acquisition/development. The suggestions were drawn from the USPACOM and subsequent projects and from research into models that were available or under development that might be useful. (These suggestions were not implemented into a Toolbox, but did form a basis for the Toolbox described in the next section.)

The short-term list included the following:

- **Unit Order of Battle (UOB) Data Access Tool**: Described above.
- **XML Tagged Scenarios**: Extensible Markup Language (XML) tagged scenarios would permit automatic use of a single scenario in multiple OOTW tools. OOTWs require scenarios to be defined down to the task level for two reasons: very low-level actions in an OOTW can have a strategic impact and our knowledge of what is important in OOTWs is so rudimentary that we must capture everything possible to avoid missing important factors. Once we know more, we will be able to determine what is unimportant and reduce the model size. Three of these scenarios were delivered by Dynamics Research Corporation (DRC) at the end of Fiscal Year (FY) 2001.
- **Scenario Viewer**: The scenario viewer is a component of situation awareness that is aimed at the analysis and planning functions. It would use an XML tagged scenario as input. It is a visualization tool that would permit the user to visually understand the various parts of the scenario.

  Figure 3.22 illustrates the concept of viewing the time and space evolution of a scenario. With a static slide, only a limited number of positions of the units can be visualized. The starting positions are at the tails of the dashed arrows. The intermediate positions are the ones shown by the unit symbols and are at the heads of the dashed arrows or tails of solid arrows. The final positions are at the heads of the solid arrows. Only limited changes can be shown without making the situation too complex to visualize. The real scenario viewer would permit viewing the evolution over time. This allows an experienced planner to look for route congestion and other time- and space-dependent events.
- **Lessons Learned Information Warehouse (LLIW) and Mine Awareness**: Described above.
- **Incidents Reporting System**: We knew some people needed an incidents reporting system—the United Kingdom (UK) built one in the field in Bosnia. It could be borrowed, modified, and put it in the hands of users quickly.
- **OOTW Toolbox Database**: Figure 3.23 illustrates a first cut at the database structure, including links, that is required to support all of the information required about OOTW tools in the Toolbox. This information includes data on the following:

Fig. 3.22 Example of a scenario viewer

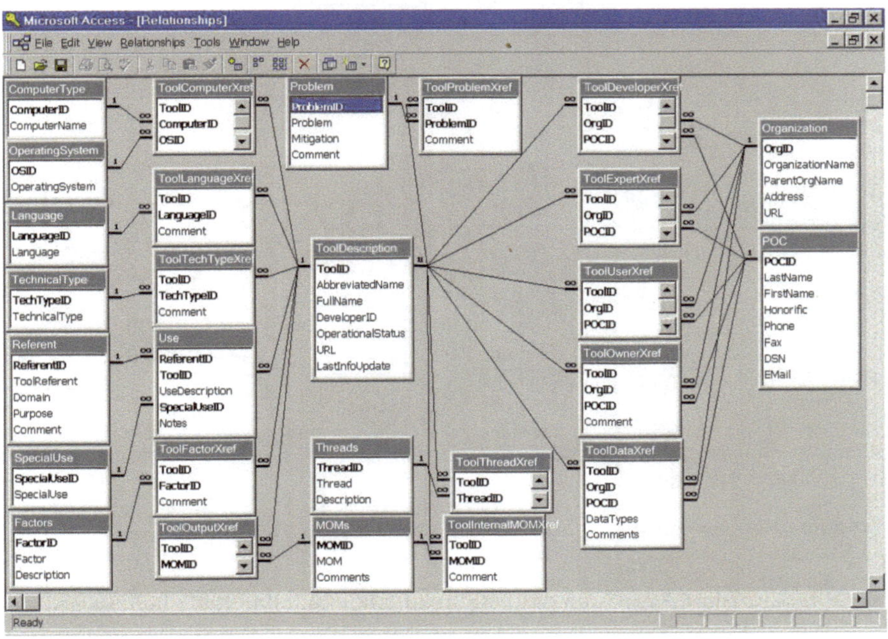

Fig. 3.23 OOTW toolbox database structure

- applications for the tool;
- purposes of the tool;
- techniques used in constructing the tool;
- factors considered, internal measures of merit, and output measures of merit;
- users, developers, owners, and experts for the tool;
- data sources for the tool;
- computers, operating systems, and languages of the tool; and
- operational status and problems of the tool.

The mid-term list included the following:

- **Pressure Point Tool:** The Pressure Point Tool is one that has received impassioned pleas for creation by personnel out of Bosnia. This tool would be aimed at the operational/tactical level. It would be used in the theater to look for potential trouble that might result from actions by the command or other actors in the area. The output consists of pages showing the local players, the neighboring players, and the international players. The labels are color-coded to show potential problem areas. When the user selects one of these and selects the factor of interest, the current distribution for that factor for that player is displayed (Fig. 3.24). (This tool was never created because it was not clear how to compute the desired results.)

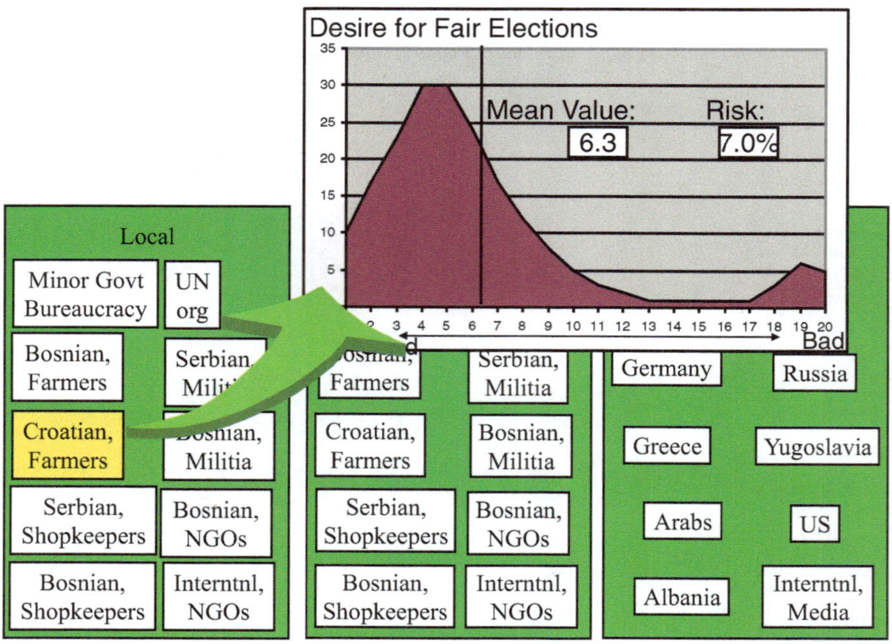

Fig. 3.24 Pressure point tool

- **Integrated Mission Planning Tool**: Mission planning is not "rocket science." However, it requires tremendous amounts of detail, with forward and backward repercussions of decisions and facts (such as availability and readiness levels). Current mission planning tools are based on the assumption that after the combat requirements are defined, non-combat requirements may be calculated by standard factors. Unfortunately, in OOTWs, the non-combat requirements are generally the drivers and the factors do not work in reverse. This leaves the planners to do everything manually. See Fig. 2.9, earlier, for a diagram of the needed model.
- **Other scenario use tools**: These include a Rules of Engagement (ROE) Viewer, a Task List Viewer, a decision support tool that uses the scenario for decision support, a force structuring tool to fit the force to the needs of the scenario, a course-of-action tool to determine the possible courses of action for a scenario, a rehearsal tool to rehearse the actions of the scenario, and a "how am I doing?" tool to determine the proper measures of merit for the scenario.

The long-term list included the following:

- **Theater Engagement Simulation**: Theater Engagement Plans Analysis is needed in each of the Combatant Commands. Theater Engagement is the Combatant Commander's contribution to "Shaping" the future; however, the Commander has no means of assessing the effectiveness of any proposed plan, nor of comparing the effectiveness of alternative plans. Some elements of this were being implemented at USPACOM in 2012 (McCurdy, 2016).

  This tool is aimed at the strategic level. It would be used by the Commander's staff (or by an Ambassador's Country Team). It is based on the concept that certain individuals are key actors, certain groups are key actors, and the concept that the bulk of the population, while important, can be modeled as a demographic/geographic distribution. Impacts spread, depending on the nature of the impact and the natures of the actors. Figure 3.25 shows a concept spreading through urban areas because of its affinity to urban populations.
- **General OOTW Simulation**: Once the Theater Engagement Simulation has been built, a more general OOTW simulation should be constructed. This general simulation can be used for COA analyses and other uses.

FAST System

In 2002, DMSO began a 4-year program implementing the OOTW Toolbox and renamed it the Flexible Asymmetric Simulation Technologies (FAST) Toolbox. Its construction was guided by the design specification as a "wish list" and a detailed review of the expressed needs of the Combatant Commands; however, considerations of what models were available and what could be afforded played a large part in its construction.

Fig. 3.25 Spreading impact

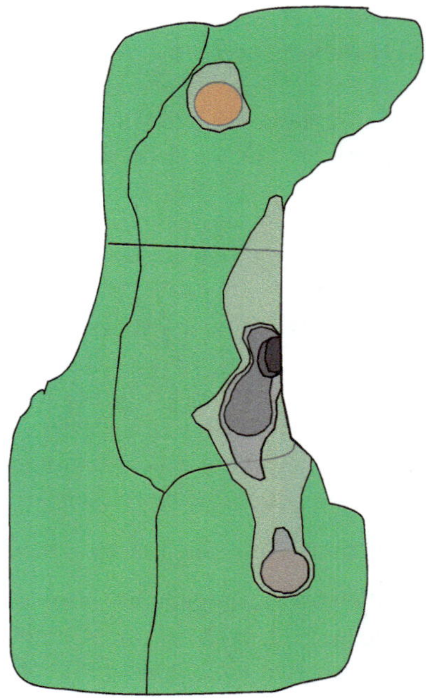

The FAST system provides an example of a system of models that together formed a model of unconventional conflict. The description of the system provides a view into many of the issues in creating an actual model of the domain and its use in a DoD study. The extensive VV&A process of the system (described in Chapter 6) provided the basis for the creation of the VV&A process for unconventional conflict and informed the ontology creation (described in Chapter 5).

The final contents of the toolbox are represented in the architecture diagram (Fig. 3.26) (Hartley, 2005a; Senko, 2005).

The tool components were the following:

- **Controller**: The controller is responsible for cataloging user data. It provides multiple views into the cataloged data, making it easier for the user to organize and share data. The controller acts as the library for the information housed within the repository. Through the controller's interface, the user can add, remove, and query content housed by the repository. The controller provides different views of the data for the user (e.g., simulation and force structure data categorized by location, application producer, and mission type).

  All application data stored in the library will be in Data Interchange Format (DIF) as defined by the family of DIFs for the FAST Toolbox. The User employs the controller to move data to and from the applications. The individual applications are responsible for the input of the data provided by the controller and the applications must output data in the appropriate DIF for movement into the repository via the controller.

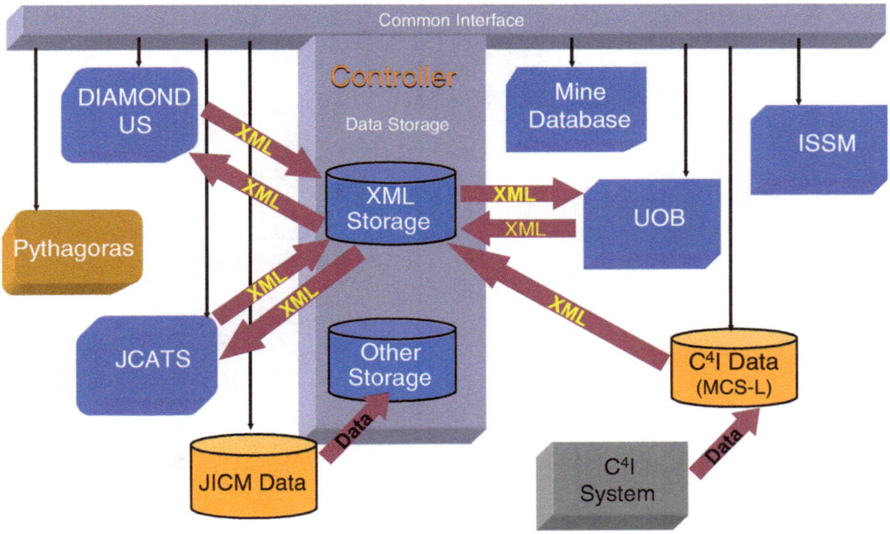

Fig. 3.26 FAST toolbox architecture

- **Canadian Forces Landmine Database (CFLD)**: Stand-alone data model, defined above, and integrated into system.
- **Unit Order of Battle (UOB) Data Access Tool (DAT)**: Stand-alone data model, defined above, and integrated into system.
- **Pythagoras**: Stand-alone ABM environment, defined above, and integrated into system.
- **Interim Semi-static Stability Model (ISSM)**: Stand-alone, DIME/PMESII, (Hartley, 2006a) dynamic (but not time-modeling) model, defined above, and integrated into system.
- **Diplomatic and Military Operations in a Non-warfighting Domain (DIAMOND)—US**: Stand-alone DIME/PMESII simulation, described above. The U.S. version of DIAMOND was integrated into system. Its scope, maturity, and usability made it the most available model for ready use in the Toolbox.
- **Joint Conflict and Tactical Simulation (JCATS)**: Stand-alone military contingency operations simulation, defined above. JCATS was selected for integration into the Toolbox because it provides a higher-resolution simulation for analysis and training. Whereas DIAMOND operates at the macro (unit) level, JCATS can operate at the detailed (individual soldier) level.
- **XPOD Manipulation Tool (XMT)**: The XMT is the utility for converting XML Populated DIFs (XPODs) between formats required by different simulations. It also extracts data from a Command, Control, Communications, Computers, and Intelligence (C4I) system to provide updated information to FAST components.

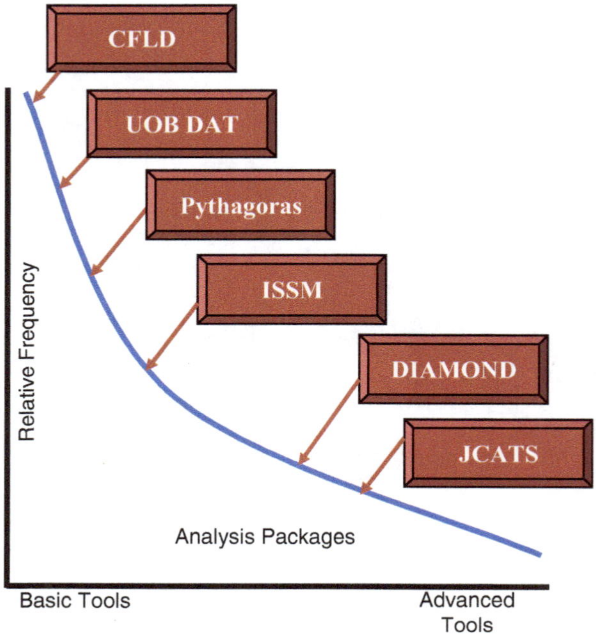

Fig. 3.27 Comparison of sophistication and frequency of FAST tools

Figure 3.27 illustrates the various levels of sophistication of the FAST tools and their expected relative frequencies of use. (Note the similarity to the canonical "Haut diagram" of Fig. 2.5.)

Table 3.3 illustrates the FAST Toolbox and individual tools utility in various military uses.

Representation of OOTW complexities requires integrated analysis across a diverse spectrum. The ISSM is the DIME/PMESII model in the Toolbox and is the primary producer of the major MOEs. It produces outputs that reflect the PMESII situation at the end of each of a number of time periods. It also computes additional changes that DIME interventions cause and computes the inferred value of "civil stability and durable peace," given the values of a set of factors within a single geopolitical area.

The ISSM accepts operational-level data modeled by DIAMOND scenarios and sub-scenarios as well as data generated by Pythagoras. It combines the data into a country-wide, strategic-level picture, displaying time-traces of the values. It also provides pre-study conversion of scenario information into the detail required to create the "Road to War" picture required for a stability operation—the detailed situation at the start of modeled events. Further, during the modeling, it provides feedback to decision-makers at the key decision points that separate the time periods to inform choices for model (e.g., DIAMOND) inputs and provides a means to calibrate the time scales used in DIAMOND for repairs. These separate time periods comprise the punctuated simulation shown in Fig. 3.9 (shown in the Modeling Constructs section above).

Table 3.3 FAST tools versus usage type matrix

V=Vital U=Useful E=Extraneous D=Distracting C=Counterproductive	Controller	XMT	UOB	JCATS	DIAMOND	ISSM	CFLD
Operations Decision Support							
COA Analysis	V	V	V	U	V	V	
Situational Awareness	U					V	
Analysis							
Deliberate Planning	U	U	V	U	V	U	
Crisis Action Planning	V	V	V	V	V	V	
Force Structure Analysis	U	V	V	U	V	U	
Acquisition Analysis	U	U	U	V	V	U	
Training							
Exercises	U	U	U	V	U		
Mission Rehearsal	V	V	V	V			
Other In-theater	U						U

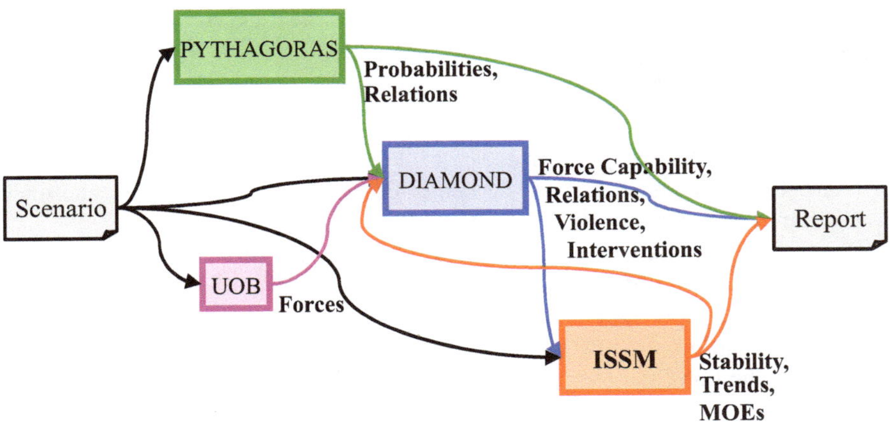

Fig. 3.28 Loosely coupled models in FAST "Battles"

The "battles" shown in Fig. 3.9 are computed in the flows shown in Fig. 3.28. The flows among the models represent the loose coupling of the models in FAST. That is, the outputs of each model are examined and modified by humans to produce the inputs to the next model. The report in Fig. 3.28 becomes the input scenario to the next cycle.

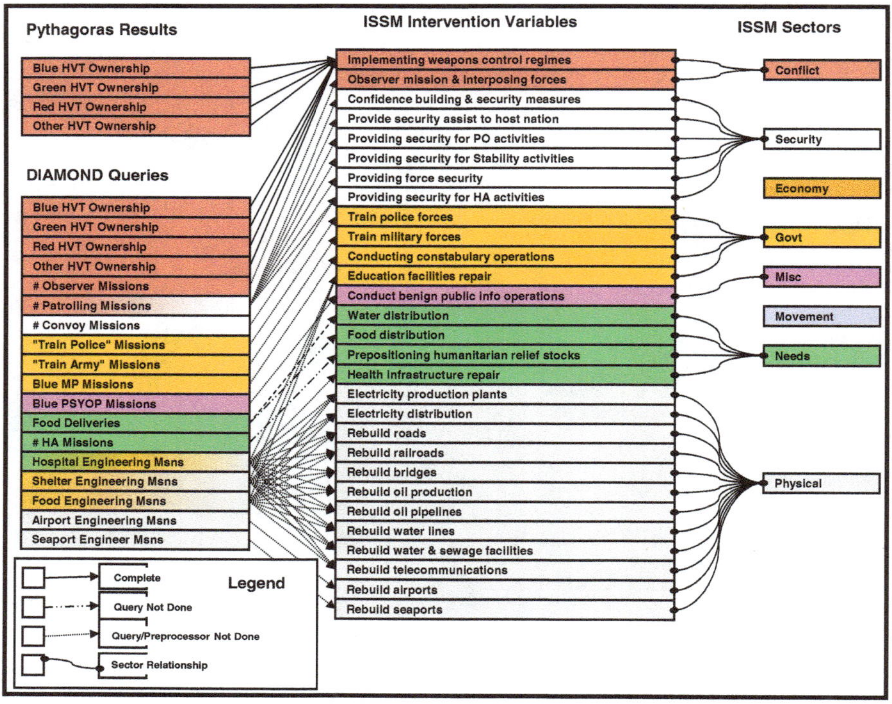

Fig. 3.29 Input sources for ISSM in the FAST toolbox

As an illustration of the loose coupling, Fig. 3.29 shows how the results from Pythagoras and DIAMOND (boxes in the left column) are converted into the standard ISSM inputs (boxes in the center column). The legend indicates that four different conversion methods were used.

The FAST Toolbox was used in a classified Enhanced Strategic Success (ESS) study for the Joint Staff in 2005. The FAST analysis portion of the study is illustrated in Fig. 3.30. This process was described in detail in Hartley, Holdsworth, and Farrell (Hartley et al., 2006).

Briefly: The ESS Study began with the study guidance and supporting materials (step 1, in the figure) and a requirement for a final report (step 14). While the analysts were developing the scenario (step 2), the UOB DAT tool was used to prepare the force lists to be used (step 3) and ISSM was used to convert the study guidance into a "road to war," which translated the text description into the PMESII variable values (step 4). The Pythagoras agent-based tool was used to develop selected (classified) inputs (step 5).

At this point, the analysis entered into a cycle of steps. Analysts created a series of synchronization matrices (step 6) and a mission analysis (step 7) from the existing situation and the campaign concepts of the commanders of the various

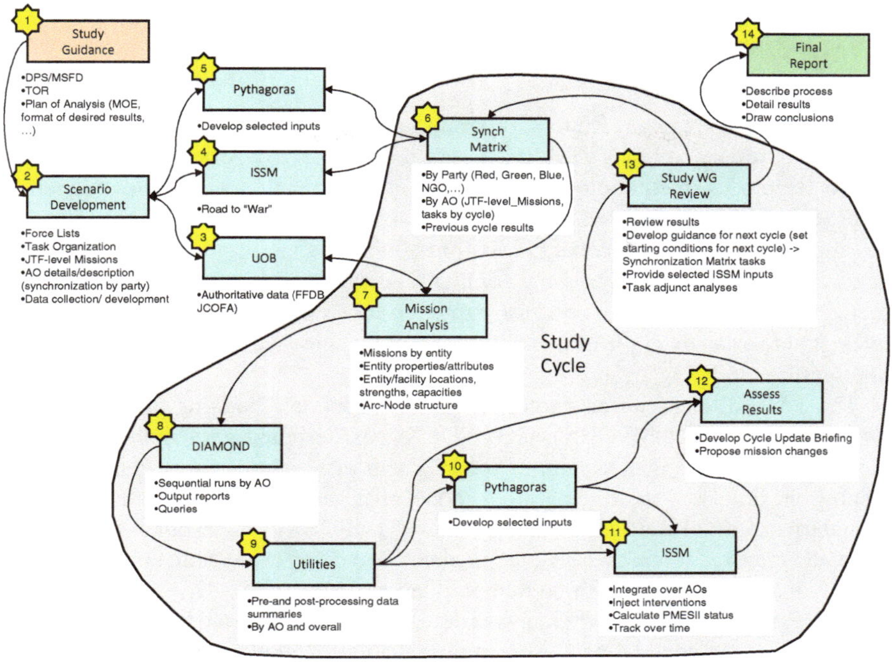

Fig. 3.30 FAST study analysis process

parties to the conflict (as articulated in the particular Defense Planning Scenario upon which the study was based). These elements were used to define the scenario for the DIAMOND simulation, which was run for a simulated period (step 8). The outputs of DIAMOND were transformed in a set of utilities (step 9) for input to ISSM (step 11) (and potentially input to Pythagoras and/or JCATS (step 10), with further use as inputs to ISSM). The outputs of ISSM and certain outputs from DIAMOND were used to assess the situation at the end of the current cycle from a modeling perspective (step 12). These results were used by the study working group to assess the situation and create the plan for the next cycle (step 13).

When sufficient cycles were completed, step 13 was followed by the creation of the final report (step 14), rather than a new cycle. In the study report, ISSM and Pythagoras produced the principal measures of merit.

To download a PowerPoint file that shows the process flow, go to http://drdeanhartley.com/HartleyConsulting/TOOLBOX/toolbox.htm and page down to the analysis process diagram, then click on the link to the file.

VV&A of the FAST Toolbox is also covered in Chapter 6.

Conflict Modeling, Planning and Outcomes Experimentation (COMPOEX) System

In late 2006 Alexander Kott, a program manager for the Defense Advanced Research Projects Agency (DARPA), asked Stuart Starr, Albert Sciarretta, and me to develop a methodology for performing verification and validation on a PMESII model DARPA was developing (Starr et al., 2007). The complexities of this model provided an excellent basis for defining the requirements and processes needed in V&V of unconventional conflict models. The verification and validation methodology will be discussed in Chapter 6; however, the PMESII model is of interest here as an example of the many parts of the unconventional conflict domain that require modeling.

The PMESII model was named Conflict Modeling, Planning and Outcomes Experimentation (COMPOEX). COMPOEX was designed as a multi-resolution model (MRM), that is, a system of models with different levels of granularity or resolution, linked so as to generate a model that accounts for the actions at fine granularity in an overall larger span model. COMPOEX was designed as a tightly coupled system of models, that is, the models comprising the system communicated directly with each other, with no human intervention.

DARPA's COMPOEX program was developing a decision aid to support leaders in designing and conducting future coalition-oriented, multi-agency, intervention campaigns employing unified actions, that is to say, a whole-of-government approach to operations. It generated a distribution of "plausible outcomes" rather than precise predictions. It provided a comprehensive family of interacting models that spanned the relevant DIME/PMESII domain. The execution of COMPOEX:

- Automatically forces models to interact to suggest plausible activities and outcomes;
- Allows the user to modify or create models and to import models;
- Allows multi-sided analysis; and
- Allows a user to visualize the interactions (normally in the form of graphical data).

COMPOEX's components included:

- **Conflict Space Tool:** This provides leaders and staff the ability to explore and map sources of instability, relationships, and centers of power to develop their theory of conflict.
- **Campaign Planning Tool:** A framework to develop, visualize, and manage a comprehensive campaign plan in a complex environment.
- **Family of Models:** These are instantiated for the current problem; however, the family concept also provided for additional models to be added to more accurately represent the operational environment.

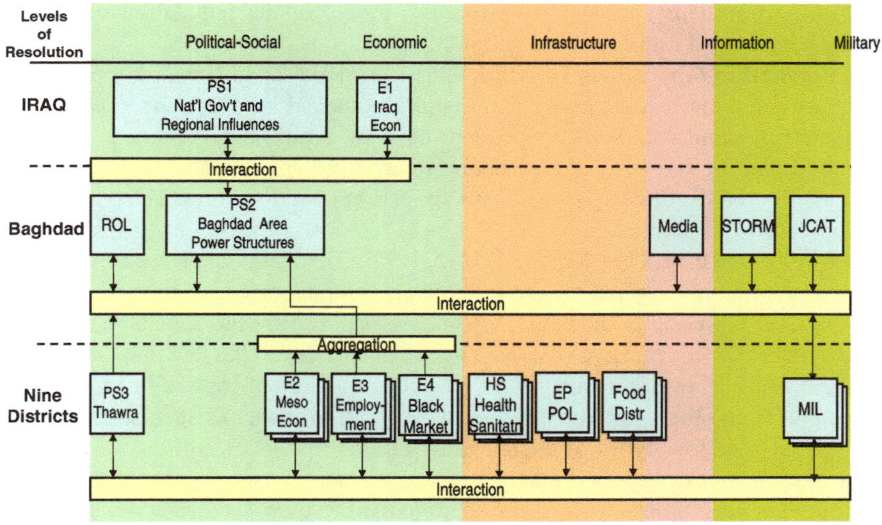

Fig. 3.31 COMPOEX model architecture

- **Option Exploration Tool:** This enables a staff to explore multiple series of actions in different environments to see the range of possible outcomes in all environments.

Figure 3.31 illustrates the structure of the COMPOEX Option Exploration Tool. Each PMESII area was represented by a set of models, divided by level of resolution. In general, the models communicated with each other by posting certain variable results to the backplane at the end of each time step and reading variable values at the beginning of the next time step. The models were tailored to the Area of Responsibility (AoR) and to the problem at hand.

Note the aggregation that occurs between the models of the nine districts and the overall Baghdad level. One critical set of models is not shown in this figure: the data that are posted to the interaction blocks (common memory) must be converted by these models to forms that are usable by other models. The aggregation step is a special case of this conversion process.

The models within COMPOEX included the following (BAE Systems, 2007):

- **EP-POL**: This is a system dynamics model of the sufficiency of the electric power—petroleum oil lubricant (EP-POL) supply: it represents stocks and flows of oil and electric power from sources to demand elements.
- **Health and Sanitation**: This is a system dynamics model of the sufficiency of the health and sanitation system (H&S). It represents stocks and flows of water (clean and soiled), and people (sick, healthy, and dead).

- **Population**: This is a systems dynamics model of the population of Iraq. It represents stocks and flows of people.
- **Information Operations (Media)**: This is a model of information operations, constructed entirely in Java code (requiring special care in connecting to the system dynamics models). It represents the flow of messages from their source to the target, the blocking of the channels carrying the messages, and the receipt and effect of the messages. Messages are part of overall themes. The output goes to STORM, described below.
- **Power Structure Tool Kit (PS1, PS2, PS3)**: This is an agent-based model of power structure. Agents can be individuals, groups, or organizations and have goals. Also present in the model are processes: these are like agents in the sense that they carry out actions, but they are not agents due to the fact that they do not have goals. They are present in the model as a source of capital, or to give/take capital from others, but are not relevant enough to model as agents.

  There are four types of capital in this model: political, military, economic, and social; links between agents are one of these types. Agents take or give capital at each time step based on their goals: however, they can only do this to other agents to which they are linked by that type of capital.
- **Rule of Law (ROL)**: This is a system dynamics model of the rule of law. It represents stocks and flows of resources and criminals.
- **Food Distribution**: This is a system dynamics model of food distribution. It represents stocks and flows of food.
- **STORM Incidents**: This is a system dynamics model of insurgent incidents. It represents stocks and flows of incidents and the capacity to affect them.
- **Military**: This is a system dynamics model of military operations. It represents stocks and flows of forces.
- **Employment**: This is a system dynamics model of employment. It represents stocks and flows of workers.
- **Macro-economy (E1)**: This is a system dynamics model of the Iraqi macro-economy. It represents stocks and flows of money.
- **Meso-economy (E2)**: This is a system dynamics model of the Iraqi meso (district level)-economy. It represents stocks and flows of money.
- **Black Market**: This is a system dynamics model of the black market economy. It represents stocks and flows of money through the black market.

While Fig. 3.31 shows how the data were passed through common memory structures, Fig. 3.32 illustrates the actual data flows among the models that comprised COMPOEX. Although all the posted data were available to all the models, each model was only interested in some of the data. This figure shows the types of data that were passed between pairs of models.

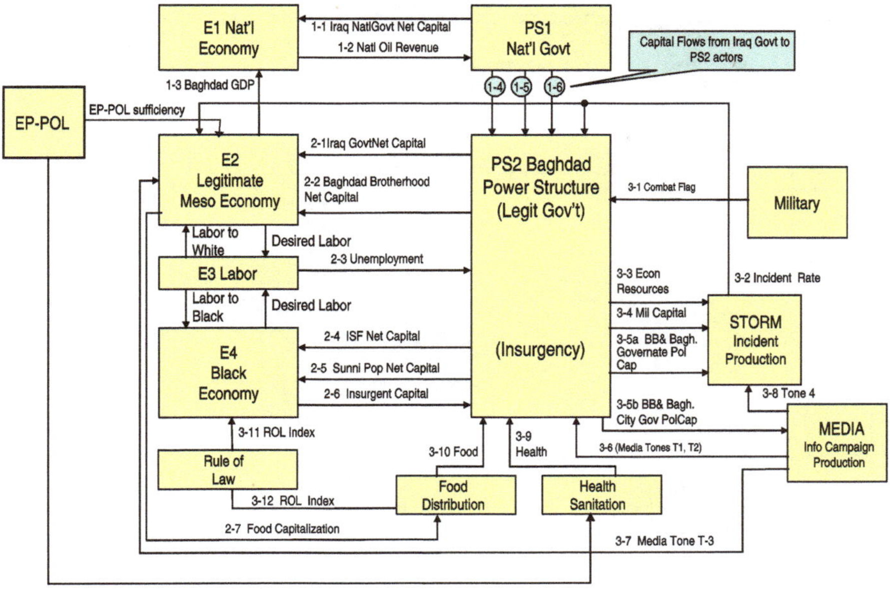

Fig. 3.32 COMPOEX data flows

HSCB Testbed

In 2009, the U.S. Army Research, Development and Engineering Command (RDECOM) Communications-Electronics Research Development and Engineering Center (CERDEC) Intelligence and Information Warfare Directorate (I2WD) at Fort Monmouth, NJ, was supporting the Human, Social, Culture, Behavior (HSCB) program. The goal of the project was to create a testbed for HSCB models. This testbed would provide the instrumentation and test facilities to examine potential HSCB models for use by DoD and provide recommendations concerning their utility, particularly in combination with other models.

Hartley Consulting became involved as the verification and validation expert. The verification and validation will be discussed in Chapter 6; however, the PMESII model and structure of the testbed is of interest here.

The testbed used a version of COMPOEX and other models, along with some additional testing structures. Figure 3.33 shows a diagram of the October 2009 demonstration system. "Ficta" was the name of the fictitious country, which was the scenario theater. The system was comprised of five components: the HSCB PMESII Model Framework (HPMF), the HPMF Server—Model Dependent, the VirtualWorld Model, the Senturion Model, and the PowerStructure Model. All of these, except for Senturion, were derived from COMPOEX. For clarity, all components at this level will be referred to as "models." The components of the models will be called "modules."

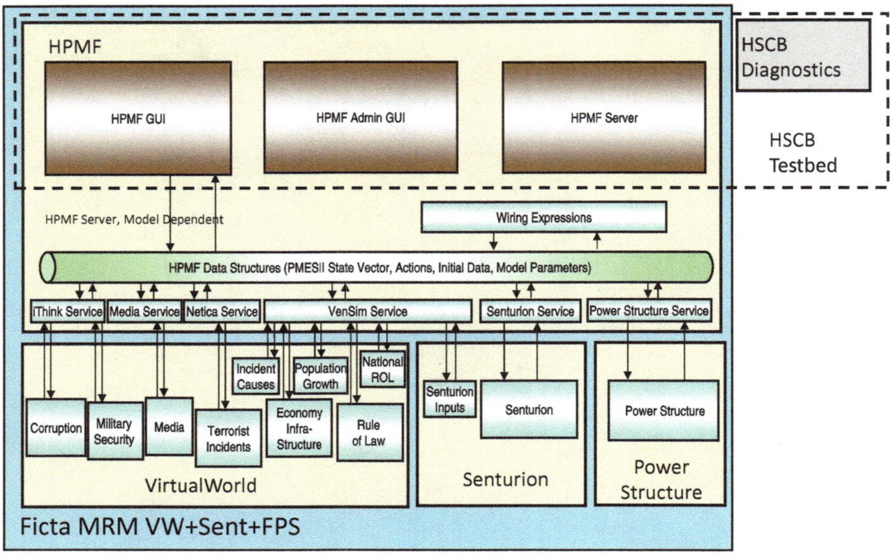

Fig. 3.33 HSCB testbed architecture

- **HPMF**: The purpose of the HPMF was to provide the common operations capabilities and integrative services for a system of models. The Framework, in its initial configuration, contained three service modules: the HPMF Graphical User Interface (GUI), the HPMF Admin GUI, and the HPMF Server.

  The HPMF GUI module provided the controls for scenario setup and exogenous time-sequenced input data for use by the system. It also provided the controls for viewing and analyzing the system results.

  The HPMF Server provided the engine to run the simulations in the system, collect and store the data, and provide other services that were broken out as "model dependent."

  The HPMF Admin GUI module supported the insertion and integration of models into a system. It also provided the controls for creating and maintaining initial data inputs for the models and other administrative tasks. Collectively, these tasks constituted the creation of the model-dependent modules, described below.
- **HPMF Server, Model Dependent**: The Model Dependent modules of the HPMF Server were divided into three categories: service modules, data structure, and modeling & simulation (M&S).

  The service modules provided the logic to translate the data flows between the models that were integrated into the system and the data structure. The system contained three generic modules that supported models based on the computer language in which they were written and three modules that supported particular named models. The three computer languages were iThink and VenSim (system dynamics programming languages) and Netica (a Bayes Net programming language). The other three modules supported a Media model, the Senturion model, and the PowerStructure model. A candidate model for inclusion in an HPMF system might be able to use one of these support modules.

The data structure module defined the data storage for the entire system. Each model had its own particular initial data and parameters, which required defined data structures and values. The set of possible actions that a user might specify in running a scenario depended on the models in the system and required defined data structures. Finally, the PMESII state vector, which provided the central integrating structure for the HPMF, required a defined data structure. The system PMESII state vector was basically the union of the sets of inputs and outputs of the models in the system. At the beginning of a time step, each model, using its service module, read the values of its inputs from the PMESII state vector. At the end of a time step, each model, using its service module, wrote the values of its outputs to the PMESII state vector. Introducing a new model into the system required adding a new segment to the PMESII state vector (among other things).

The final Model Dependent module was a custom model, called the HPMF Wiring Expressions module. This module consisted of a set of statements that connected the data slots of the PMESII state vector. For example, suppose a module computed population growth during a 1-week time step. It began by reading the starting population size from the appropriate slot, computed the increase for the week, and finished by writing the ending population size to an appropriate output slot, which was different from the input slot. For the next time step, the population should not start from the same value as it did in the prior time step, it should start with the ending value from the prior step. A simple Wiring Expression entry would take care of that:

"At the end of a time step, replace the population input value with the population output value."

The system would evaluate population growth for 1 week, output the result, start with that output and evaluate growth for another week, output the result, and continue for the desired length of time. If all that was desired was a population growth model, the Wiring Expression entry would be contained in the model; however, exposing it allowed for integration with other models. For example, suppose there was another model that reduced population, e.g., by terrorist events or by natural disasters, outputting deaths at the end of each time step. A more complex Wiring Expression entry would combine the two:

"At the end of a time step, replace the population input value with the population output value minus deaths."

Thus the PMESII state vector was the central integrating structure of the HPMF and the Wiring Expressions module was the central integrating mechanism for the HPMF.

- **VirtualWorld Model**: The VirtualWorld model contained major modules for the economy and infrastructure, terrorist incidents, rule of law, corruption, media, and military security. It also contained three smaller modules to help with incident causes, population growth, and national rule of law. This model provided the basic virtual world of the scenario.
- **Senturion Model**: Senturion is a stand-alone simulation, described above, that was integrated into the testbed. This installation also included a small module to

help convert outputs from other modeling constructs into inputs suitable for the stand-alone Senturion model. The stand-alone model was the other module in the integrated Senturion model. Once Senturion has run, the Senturion time stream had to be merged with the overall model time stream.

- **PowerStructure Model**: The PowerStructure model simulated multiple factions and their power struggles within a scenario. It also was its own module.
- **FactionSim Model**: FactionSim was a stand-alone simulation, described above. It was planned to be part of the system; however, last-minute problems in converting ABM outputs into inputs for system dynamics models prevented this (Waltz & Lawrence, 2009).

One of the lessons that was learned from the HSCB Testbed was that the types of models that might address unconventional conflict modeling were too various to be supported by a single testbed.

Modeling Requirements

The Center for Technology and National Security Policy at the National Defense University (NDU) held a series of workshops on HSCB modeling in 2008 and 2009. The results of the first workshop were published (Woodcock, Baranick, & Sciarretta, 2010).

One of the major results of the first workshop was a categorization of the capabilities needed for HSCB modeling. Figure 3.34 shows the linkages of the categories. It shows the flow from "representative questions" that drive the

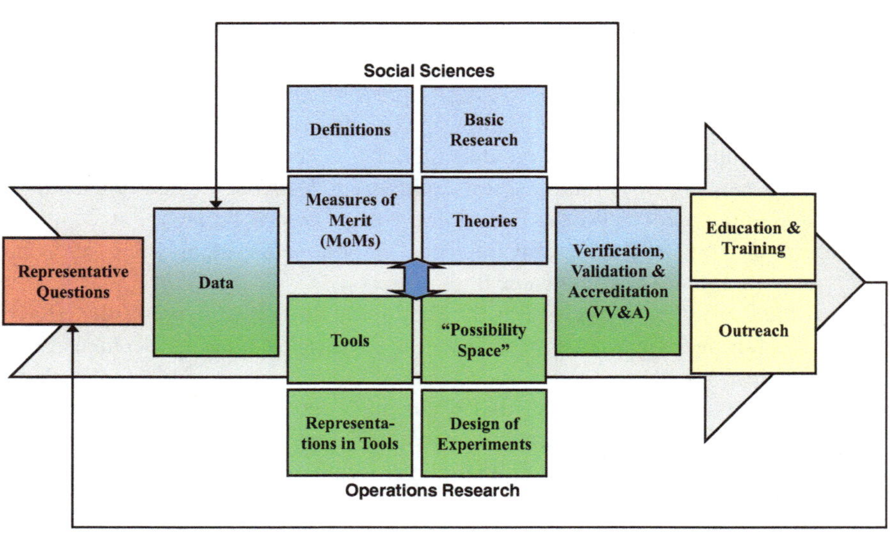

Fig. 3.34 Capabilities needed for HSCB modeling

modeling needs to “education and training” and “outreach” to socialize the results. It also shows the pairing of the “social sciences” categories with the “operations research” categories and the link back to “data” required for the “VV&A” category. Each category is discussed below.

Representative Questions

The question of what capabilities are needed starts with defining the questions that drive the modeling process. A representative selection of these questions follows:

- What are the forecasts for, among other things, refugee flows, contagion of disease, authenticity of cultural change?
- How can the generation of Theater Security Cooperation Plans (TSCPs) and situational awareness be supported?
- How can the flow through the Stability, Security, Transition, and Reconstruction (SSTR) process be balanced (e.g., integrity of host nation decisions while supporting U.S. interests)?
- How should options be prioritized?
- What SSTR operations should be undertaken?
- What Science and Technology (S&T) investments are required?
- How can we understand such things as failed states, legitimacy, deterrence/influence/containment, why people become insurgents, unintended consequences, governance?

Data

Existing HSCB data sets are diffused and difficult to find and access. They lack necessary information to support analysis (e.g., adequate metadata and indications of pedigree). And they are rarely ready for use—they require cleanup and conversion to fit current needs. Suggested actions were to:

- Develop appropriate HSCB lists, taxonomies, and ontologies;
- Implement efforts to tailor HSCB data to satisfy the intended purposes;
- Perform and record data V&V efforts (e.g., integrity, consistency, reliability, and source) as metadata;
- Update local and national data, with appropriate periodicity;
- Capture data capabilities in many dimensions (e.g., environmental, medical, attitudes, affiliations, and legal systems); and
- Perform an assessment of the desirability of a central HSCB data repository (with associated issues of classification, access, open source data, legal, granularity, qualitative data, maintenance, and dissemination).

Definitions

Many HSCB-related definitions are too vague or ambiguous (e.g., "counterinsurgency"). And many definitions are discipline-specific and not widely understood. The two recommendations were to develop more specific definitions to drive action and develop a variety of products including a library of ontologies, semantic descriptions, thesauri, dictionaries, data lexicons with metadata, and standards.

Measures of Merit (MoMs)

The community does a poor job in formulating Measures of Merit (MoMs) for HSCB and in developing the relationships (cause and effect) that link them. MoMs include the following:

- Dimensional Parameters (DP): Properties or characteristics inherent in the physical entities, e.g., production rate of a water purification unit.
- Measures of Performance (MoP): Measures of direct results. Examples are miles of road constructed, number of bridges built, and increases in crop production.
- Measures of Effectiveness (MOE): Measures of effects that depend on purpose, such as popular satisfaction with an agricultural program.
- Measures of Force Effectiveness (MoFE): MOEs that relate to an entire organization's activities, such as pacification.
- Measures of Political Effectiveness (MoPE): MOEs that relate to the effectiveness at the highest level, such as civil stability and durable peace.

The suggested actions were to tailor the MoMs to HSCB problems of interest and develop relationships that link them and display HSCB MoMs to operational users and senior decision-makers in a fashion that conveys appropriate levels of uncertainty and risk.

Theories

In selected areas of social sciences, there are excessive numbers of theories (e.g., "the root causes of terrorism"). However, there are key areas where needed theories are lacking, too particular, or too general. When multiple theories exist, the community needs codes of best practice/guidelines on which theory to use and when. There is also a need to develop appropriate social sciences theories to address key gaps or mismatches.

Social science has two distinct regimes: micro-theory and macro-theory. Micro-theory deals with individuals and macro-theory deals with the society. Ideally, these should be connected or at least connectable. Kuznar discussed the hope that macro-theory would be an emergent property of micro-theory and described how modeling may lead to this discovery (Kuznar, 2010).

Loren Cobb discussed some failures and a success in the applications of macro-theory to modeling (Cobb, 2010). "In models of refugee flow during political crises, as in the Rwandan terror campaign of 1994, for example, existing social theory proved to be almost useless in predicting when people would leave their homes, in which direction they would decide to go, and where they would decide to stop." "In social epidemiology, in contrast, the mathematical theory developed in the twentieth century does a reasonably good job of predicting the flow of most contagions through the population-density map of contemporary societies of all types." "On the other hand, the intergenerational social epidemiology of post-traumatic stress disorder (PTSD) reveals a serious gap in social theory." Social theory is silent on the impact on society of epidemic levels of PTSD over multiple generations.

Jack Jackson discussed a multi-agent systems model of civilian populations in reconstruction operations (Jackson, 2010). The model was built to represent the flow of information, the beliefs and positions of agents, the importance of persuasion, and the theory of planned behavior.

Basic Research

Many critical HSCB areas are not well understood at the social sciences level. Among others, the HSCB community needs research on:

- Interior cognitive models;
- Influencers on attitude/behavior of civilians based on ethnic, tribal, cultural, religious, and political considerations;
- Local legitimacy; and
- Trust.

Representations in Tools

The entire DIME/PMESII spectrum needs to be represented; however, each individual tool need not cover the entire spectrum. The community needs to:

- Model ourselves as well as "others";
- Provide feedback between "micro" and "macro" representations; and
- Capture organizational performance, cultures and institutions, all types of operations, and situational awareness of all parties.

Tools

There is no such thing as a "silver bullet" model that will satisfy all of our HSCB modeling needs. Many existing HSCB tools are too limited in their scope (e.g., counterinsurgency tools cannot address multiple insurgencies, simultaneously). And the community must understand the limits to forecasting when addressing HSCB issues.

- There is a need for an HSCB modeling test bed to support multiple functions (e.g., evaluation, transition).
- The proper architecture/framework for such an HSCB modeling test bed is an open question; however, there are many suggestions:
  - The suite of tools should include a balanced mix of techniques that take advantage of the strengths of the tools while ameliorating their weaknesses (e.g., system dynamic models; agent based models; wargames);
  - Consideration should be given to creating a "collaborative environment" in which selected models can be linked/federated and evaluated (particularly with respect to "precision");
  - Consider the use of a "service bus" for data repositories;
  - Ensure that models are tailorable;
  - Employ hierarchical modeling with meta-model/meta-data aggregation/dis-aggregation; and
  - Assemble a resource repository of models and data.
- The experience with the HSCB Testbed (described in the previous section) showed some of the difficulties in implementing this concept.

"Possibility Space"

Not all models need to "forecast" things (see representative questions, above). "Forecasting" may often mean likelihoods of occurrence of multiple possible events. Thus, rather than using the term "forecast," we might speak of a "possibility space" that includes (in different situations):

- "Hard" prediction of events;
- "Soft" prediction of likelihoods;
- Exploration of possibilities; and
- Situational awareness and understanding.

Design of Experiments

Many users of HSCB M&S do not understand how to derive statistically meaningful insights from their tools.

The HSCB community should draw on the insights developed in academia (e.g., at NPS for M&S characterized by large numbers of variables) to characterize the interesting parts of response surfaces using innovative experimental designs (e.g., focused fractional factorial designs and nearly orthogonal experimental designs) (Cioppa & Lucas, 2007).

Verification, Validation & Accreditation (VV&A)

Most creators of HSCB M&S and selectors of data do not do an adequate job of V&V. The HSCB community needs to broaden the concept of V&V to theories and selection of Subject Matter Experts (SMEs). It is important to initiate the VV&A process even when key issues are still unresolved (e.g., precise definitions, levels of accomplishment, descriptions of pitfalls, etc.). The community needs to understand that V&V of DIME/PMESII data is significantly more complex than V&V of combat model data. Suggested actions were to:

- Generate guidelines that enable us to perform V&V (for theories, tools, data, and SMEs) credibly, with acceptable levels of resources;
- Pay additional attention to the accreditation function;
- Use V&V to create "tags" for theories, methods, models, and data to allow retrieval of desired item when needed; and
- Create open data on detailed VV&A status of models and data.

Some of these needs are addressed in Chapter 6.

Education and Training

It will take high-performance, compatible, multidisciplinary teams (e.g., social scientists, operations analysts) to create and employ HSCB modeling. However, those diverse communities have a difficult time in communicating. It is even more difficult to communicate HSCB assumptions and results to decision-makers than for combat models. And it is extremely important and difficult to "train as we fight" in a HSCB environment. The community needs to:

- Augment the curriculum for social scientists and operations analysts to give them adequate education to enhance cross-discipline communication;

- Create and sustain a HSCB Community of Interest (COI) (perhaps drawing on MORS Social Sciences COI) to foster high-performance, multidisciplinary teams; and
- Develop the tools and data needed to "train as we fight" to support both the education and training and operational communities.

Outreach

The HSCB community must include interagency participants, not just the military. Further, HSCB modeling is of interest to our international partners. However, the community must realize that some members of the social sciences community have argued that it would be "unethical" to collaborate with DoD. The community needs to:

- Expand the HSCB COI to include balanced interagency participants (e.g., National Security Council (NSC), United States Institute of Peace (USIP), Department of State (DOS), Department of Justice (DOJ));
- Participate in international forums that address HSCB modeling issues (e.g., North Atlantic Treaty Organization (NATO) System Analysis and Studies (SAS) initiatives on HSCB, Irregular warfare); and
- Develop a compelling narrative to explain the value of a collaborative relationship between the social sciences community and DoD.

Chapter 4
The DIME/PMESII Paradigm

A good paradigm is worth ten parapennies.

There was a time when the phrase "paradigm shift" was popular—and overused. A true paradigm shift occurs when a new thought pattern is so useful that the previous conceptual world view is replaced by a new one. The DIME/PMESII paradigm did just that. In this chapter, we explore the central concepts of DIME/ PMESII, its variants, and its criticisms.

The origins of the DIME/PMESII paradigm are unclear. There have been many articles written about various versions of the paradigm, but none seem to have cited an origin. Probably, it has developed through the joint efforts of the unconventional conflict community, with no individual creator.

PMES Concept

Figure 4.1 was included in a presentation I created in late 2001 for a presentation at George Mason University (GMU) in 2002. I don't remember whether I created the diagram or modified it from some other source. However, it clearly represents a first step in developing the PMESII concept. It shows the interactions among the Political, Military, Economic, and Social (PMES) state variables within each of two countries and the interactions between the corresponding state variables of the countries.

PMESII Concept

The PMES concept became the PMESII concept sometime before 2006, as shown in Fig. 3.6. Figure 4.2, taken from the U.S. Joint Publication 3-0 (Chairman of the Joint Chiefs of Staff, 2011), shows a different version of the figure illustrating the

D.S. Hartley III, *Unconventional Conflict*, Understanding Complex Systems,
DOI 10.1007/978-3-319-51935-7_4

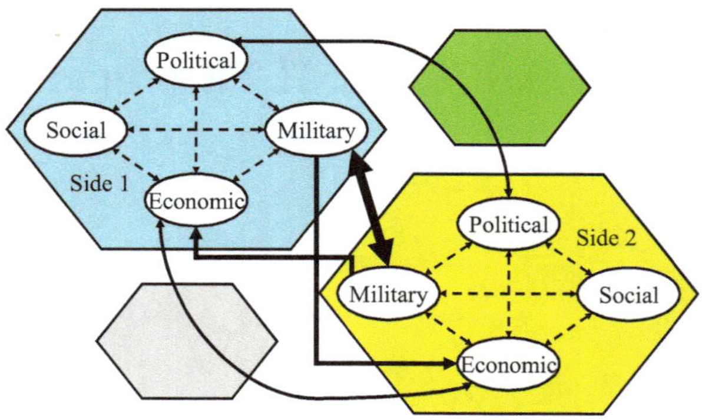

Fig. 4.1 Military conflict modeling PMES diagram

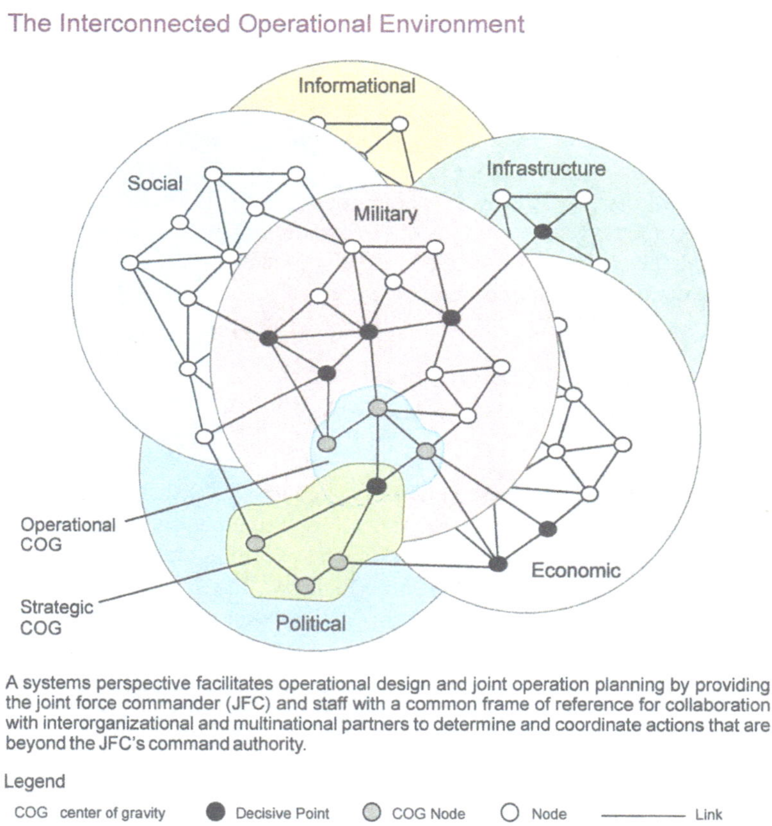

Fig. 4.2 PMESII diagram

same point. The five domains, political, military, economic, social, information, and infrastructure, are shown as parts of the operational environment. Each has its own link-and-node structure; however, all are also interconnected. The figure emphasizes that the operational center of gravity (COG) may differ from the strategic COG.

Variations to PMESII have been proposed, such as PMESII-PT, which adds physical environment and time to the package. However, the central PMESII construct has proved to be durable and useful and that term will generally be used to refer to all variants.

Alexander Kott and Gary Citrenbaum produced and edited a book with chapters that elucidate many of the details behind the PMESII construct (Kott & Citrenbaum, 2010). Selected chapters are cited in the following text.

Political

The primary components of the political domain are governance (policies, personnel, organizations, freedom, etc.), the rule of law (judiciary, law enforcement, crime, etc.), and politics (leadership, factions, etc.), as well as some miscellaneous items (intervenor status, stability/peace and legitimacy ratings, etc.). Measuring the status of some of these items is difficult but obviously necessary.

Abdollahian, Kugler, Nicholson, and Oh discussed politics and power (Abdollahian, Kugler, Nicholson & Oh, 2010). Corey Lofdahl discussed governance and society (Lofdahl, 2010). Duong, Turner, and Selke discussed crime and corruption (Duong, Turner, & Selke, 2010).

Military

The primary components of the military domain are conflict, government (organizational sizes, intelligence services, the relationship with the government, etc.), and security (provisions), as well as other items (insurgents, terrorists, paramilitary forces, capacities, etc.). Measuring the status of these items is easier than measuring the political status.

Abdollahian, Kugler, Nicholson, and Oh discussed politics and power (Abdollahian et al., 2010). Kott and Skarin discussed insurgency and security (Kott & Skarin, 2010).

Economic

The primary components of the economic domain are agriculture, crime, energy, finance, governmental economic actions, jobs, and other things. Measuring the status of economic items is a relatively well-defined process compared to some other PMESII items.

Alan Graham discussed economics and markets (Graham, 2010). Duong, Turner, and Selke discussed crime and corruption (Duong et al., 2010).

Social

The primary components of the social domain are basic needs (food, water, shelter, etc.), education, health, movement (freedom of, restrictions on, forced, etc.), safety, and other items. Some of the items are relatively easy to measure, while some of the "other items" that include religious and associational metrics are quite difficult to measure.

Bhavnani, Miodownik, and Riolo discussed groups and violence (Bhavnani, Miodownik, & Riolo, 2010). Duong, Turner, and Selke discussed crime and corruption (Duong et al., 2010).

In 2013, Gregory, Engel, Bell, and McGrath discussed methods they developed "to identify online communities, or groups, using a combination of structural information variables and content information variables from weblog posts and their comments to build characteristic footprints for groups" (Gregory, Engel, Bell, & McGrath, 2013). They found "that these variables do a good job at identifying groups, placing members within a group, and helping determine the appropriate granularity for group boundaries."

Informational

The primary components of the informational domain are general information items (primarily decision-making ratings), media (capacity, freedom, etc.), opinions (of various actors about legitimacy, satisfaction, etc.), and information operations.

William Bennett discussed the media and influence (Bennett, 2010).

In 2013, Pfeffer and Carley analyzed structural change in Arabic written data and compared it to traditional English written news sources in Syria in December of 2011 (Pfeffer & Carley, 2013). They were "able to show that online and offline data is significantly correlated but the change can be measured earlier in online networks." There was also a distinct connection between real-world activities and the information collected, but that "social media is a coordination mechanism that enables and mitigates activity rather than causing activity."

Infrastructure

The primary components of the infrastructure domain are business infrastructure, social infrastructure, energy infrastructure, government infrastructure, transportation infrastructure, and water infrastructure. There are a very large number of items that fall into this domain, including capacities, damage, investments, and so on.

In 2010, Alan Graham discussed economics and markets, with some connections to infrastructure (Graham, 2010). Corey Lofdahl discussed governance and society with some connections to infrastructure (Lofdahl, 2010).

DIME Concept

The DIME concept rests on the observation that a nation has many instruments of power with which it can attempt to achieve its international goals. These instruments or levers of power can be collected into a small set of categories to make remembering them easier. The most popular set is DIME—Diplomatic, Information, Military, and Economic.

Diplomatic power rests in negotiations and agreements. To the extent that a nation regards itself as bound to honor its agreements, diplomacy can result in changes in the actions of one or more nations. However, there are advantages to be had short of binding agreements. Offering to provide or withhold favors among countries can also result in changes of actions.

Information power lies in gaining information from others and in controlling the information desired by others. Differences in what is known between contending parties can be decisive, whether the domain is economic, military, or diplomatic.

Military power is an obvious component. However, the fact that it has limits to its effectiveness and on its use should serve warning that the other levers of power also have limits.

Economic power is also an obvious component. Nearly immediate effects can be seen with such things as freezing bank accounts. Long-term strategies involve the stronger economic power causing the weaker power to spend itself into defeat.

Variations to DIME have been proposed, such as DIMEFIL, which adds Financial, Intelligence, and Law Enforcement to the DIME package.

Unified DIME/PMESII+ Paradigm

As mentioned earlier, there have been variations of DIME/PMESII, which have been argued (including my offering, adding Kinetics and Environment as state variables); however, for our purposes, we will use PMESII to refer to all state vector variables, regardless of taxonomy details. The acronym DIME refers to the levers of power that a (nation) state has to influence the PMESII state. As with PMESII, we will use DIME to refer to all such interventions, regardless of taxonomy details. Collectively, these will be referred to as DIME/PMESII+ or DIME/PMESII or simply as PMESII for brevity (Hartley, 2015a).

Table 4.1 lists my PMESII subcategories as an aid to understanding the breadth of the paradigm.

Table 4.1 PMESII subcategories

Category	PMESII subcategory
Political	Political-Gov
	Political-Pol
	Political-ROL
	Political-Overview
Military	Military-Confl
	Military-Gov
	Military-Sec
	Military-Oth
Economic	Economic-Agri
	Economic-Crime
	Economic-Energy
	Economic-Financ
	Economic-Gov
	Economic-Jobs
	Economic-Oth
Social	Social-BasicNds
	Social-Edu
	Social-Health
	Social-Move
	Social-Safety
	Social-Oth
Information	Info-General
	Info-Media
	Info-Opinions
	Info-Ops
Infrastructure	Infrastructure-Busin
	Infrastructure-Social
	Infrastructure-Energy
	Infrastructure-Gov
	Infrastructure-Transport
	Infrastructure-Water
Kinetics	Logistics
	Damage/Attrition
	Kinetics-General
	C4I
Environmental	Atmospherics
	Geometrics
	Earth-water-other

Below the subcategories are the classes that comprise the heart of the paradigm. Table 4.2 lists a few of these classes. (Note that there is a naming convention for the classes that is rational but not in Standard English language style manuals.) To download a file containing all of the categories, subcategories, and PMESII+

Table 4.2 Sample PMESII classes

Category	Subcategory	ClassName	Defining questions
Political	Political-Gov	*GovtDecisionAuthorityRating*	Who has power (autocratic, democratic, theocratic, monarchy, warlords, etc.)?
		GovernanceRating	How well does it govern; what are the checks on its power?
		FirstRespondersActivityRating	How many first responders are there; what is their influence and activity?
		GovtLeaderChangeRating	How does leadership change: elections, hereditary succession, coup, revolution?
		CentralAuthorityEffectivenessRating	How effectively does it govern: political capacity (tax collection capability, etc.), influence events?
		SocialServicesAdequacyRating	How adequate are child services, elderly care, etc., in scope (distribution throughout society) and effectiveness?
		GovtDecisionMakingRating	What is the distribution of power, number of political parties, checks and balances, elections, dispute resolution?
		RelationshipWithIntervenorsRating	What is the relationship between the parties, ranging from adversarial to very close?
		FreedomRating	What is the freedom rating (use something like the FreedomHouse scale)?
		ExternalGovtAdvisorsMoP	What is the number and influence of external advisors to the HN government (not counting Rule of Law advisors)?

classes go to http://drdeanhartley.com/HartleyConsulting/IWMetricOntology/IWOntology2.htm and click on the download link in the description to the left of the metric ontology figure.

Beyond PMESII

Discussions about DIME/PMESII continue, indicating that the concept, while providing a useful construct, is not completely satisfactory.

In addition to alternatives to PMESII, there are alternatives to DIME. Recent publications use Military, Information/Intelligence, Diplomatic, Legal, Infrastructure, Finance, and Economic or, alternatively, Military, Informational, Diplomatic, Law Enforcement, Intelligence, Financial, and Economic (MIDLIFE) as a replacement for the DIME concept of elements of national power (Burke & Self, 2008).

Further, there are arguments against the basic construct. Maj Brian Ducote, U.S. Army, has argued that the PMESII concept, in its application, is not advancing the goal of increasing the understanding of unconventional conflicts (Ducote, 2001). He argued that the linear structure of the PMESII concept only reveals the "what," but not the "why," of complex systems. He advocated that a more ontological and holistic approach is required to achieve a deeper understanding of unconventional conflicts.

The next chapter (Chapter 5), which concerns an ontological representation of the general PMESII paradigm, addresses some of these concerns, although it may not satisfy all of them.

Chapter 5
Ontologies

A branch of metaphysics concerned with the nature and relations of being.

In this chapter, we are not discussing metaphysics, rather we are interested in computer science ontologies. According to Wikipedia the definition of this is almost equally baffling: "an ontology is a formal naming and definition of the types, properties, and interrelationships of entities that really or fundamentally exist for a particular domain of discourse" (Ontology (information science) n.d.). In this chapter, we provide a more extensive and concrete definition of ontologies that should make the meaning clear. We then proceed to describe an ontology for unconventional conflict.

Ontology Definition

Figure 5.1 begins the definition of an ontology by showing types of domain knowledge that might be contained in an ontology.

For example, lists of terms or lists of terms with definitions (glossaries) are placed toward the "informal" side of the figure. Thesauri and lists of subterms are also on the "informal" side, but closer to the "formal" side. The dashed, vertical line indicates a common division line between things that are not ontologies (informal) and things that are ontologies (formal).

Lists and definitions certainly contain information about the domain. A definition is a property that restricts a named class (Fig. 5.2). An individual instance (instantiation) of that class must meet the terms of that definition and provides a value for the property.

Thesauri add a relation among elements that is similar to a definition, but looser—more general. Terms are linked that do not have the same definition, but have similarities of definition. Subterms also have a generalization of the definition property, but one that consists of a restriction of the definition, with a different restriction for each subterm.

D.S. Hartley III, *Unconventional Conflict*, Understanding Complex Systems,
DOI 10.1007/978-3-319-51935-7_5

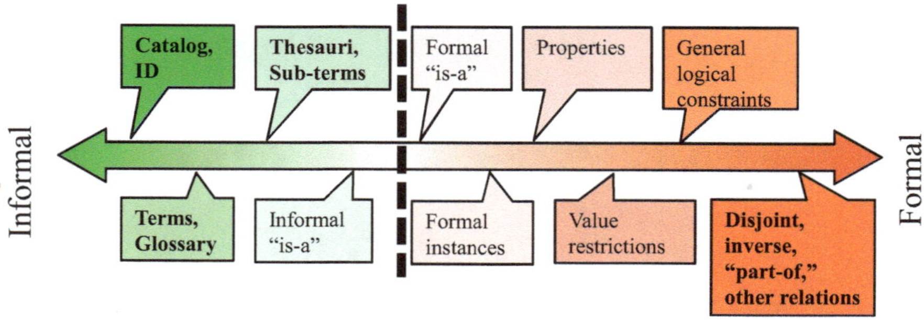

Fig. 5.1 What is an ontology?

Fig. 5.2 Information representation

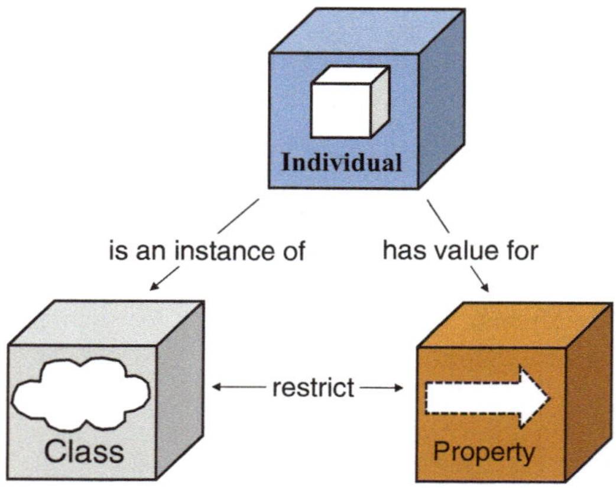

Fig. 5.3 "Is-a" relation

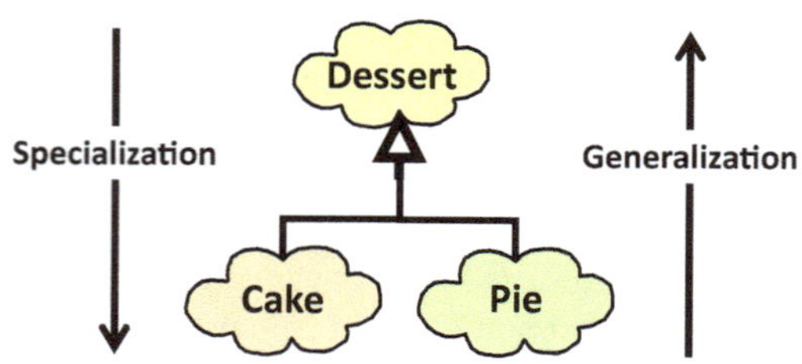

The next stage of formality is the introduction of other relations between classes. The first such relation is the "is-a" relation, which is the standard relation in many taxonomies. Figure 5.3 illustrates this by showing that the class "*Pie*" "is-a" "*Dessert*" and the class "*Cake*" also "is-a" "*Dessert*." The class "*Dessert*" has two child classes; however, each of these has only one parent class. This is a property of taxonomies, a type of ontology. Notice that once you get to the stage of drawing taxonomies, you have crossed over from informal "is-a" to formal "is-a."

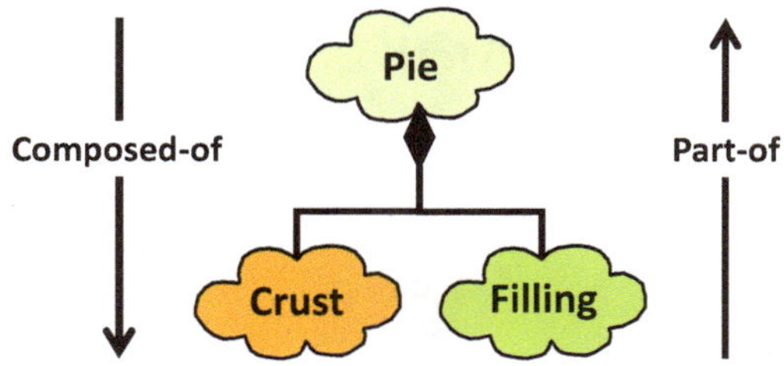

Fig. 5.4 "Part-of" relation

The next stage of information capture in an ontology is the formal distinction between classes and instances. This is different from the further decomposition of the taxonomy in Fig. 5.3. The taxonomy decomposition might result in defining a "*CherryPie*" class and an "*ApplePie*" class, each a child of the "*Pie*" class. An instance of the "*Pie*" class would be a particular pie, so that a bakery might have five instances of pies in its display case.

The next stage shown in Fig. 5.1 is labeled "properties." You may have noticed apples in the grocery store with labels on them, each label having a number. Winesap apples might all have the number 4189 and Granny Smith apples might all have the number 4139. These are price look up (PLU) numbers. In the fruits taxonomy, the parent class would be "*Fruit*," one child class would be "*Apple*," which would have child classes "*Winesap*" and "*GrannySmith*." At this lowest level, the classes would have a property "*PLU*." The value of PLU would depend on the class. This is an extension of the definition-property concept.

At the farthest "formal" end of Fig. 5.1, there are a set of additional information types. One of these is "part-of," which is a relation between classes, as is "is-a," but differs from it in the nature of the relation. Figure 5.4 illustrates this relation.

The "*Pie*" class can also be described by what it is composed of. Here it is shown as being composed of a "*Crust*" class and a "*Filling*" class. (Reading the diagram in the opposite direction, the "*Crust*" class is "part-of" the "*Pie*" class, as is the "*Filling*" class.

Notice that this diagram is also a taxonomy; however, if you want to display everything you know about the "*Dessert*" domain, you now have two separate taxonomies. If you want to combine them, you no longer have a taxonomy because there are two relations being displayed. Now you have an ontology that is not a taxonomy.

To complete this discussion, suppose that the overall domain is not "*Dessert*" but is "*Food*." Somewhere in the domain is the "*Pizza*" class. "*Pizza*" "is-a" "*Pie*," but "*Pizza*" is not a "*Dessert*." This means that the class "*Pie*" has to have a parent, such as "*MainCourse*," as well as having "*Dessert*" as a parent. Taxonomies are not allowed to have multiple parents for a child; however, the more general ontology does not have this restriction.

In general, ontologies allow you to list and define the concepts of a domain and to describe all of the relations that exist among the concepts. They also permit distinguishing between the concept classes and the instances (also known as objects) of the classes.

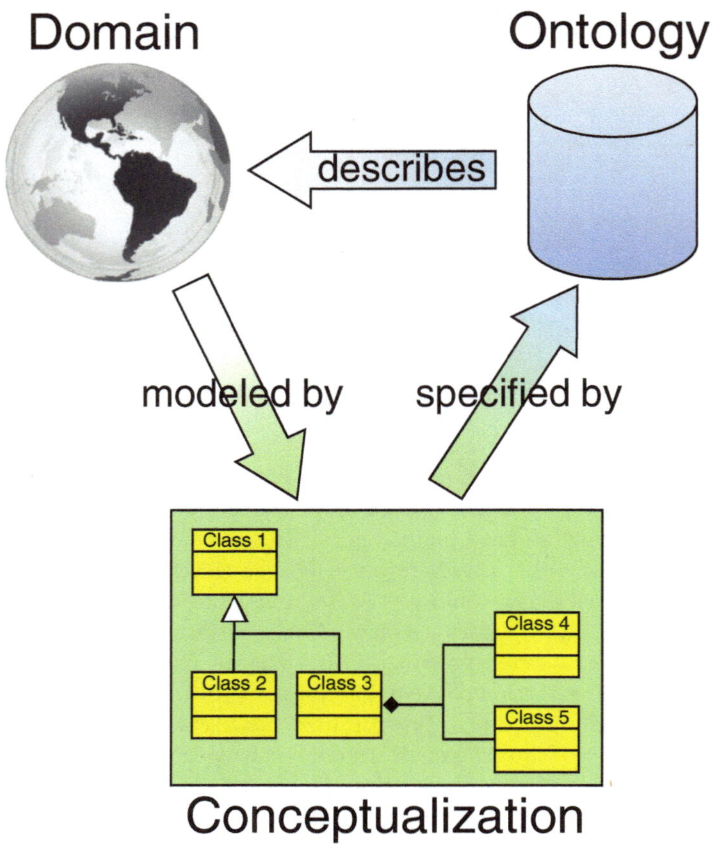

Fig. 5.5 Domain, ontology, and conceptualization

Figure 5.5 illustrates how this all fits together. The ontology describes the domain. It specifies one or more conceptualizations—techniques for visualizing the ontology—such as a class relationship diagram. This conceptualization is a model of the domain.

In this figure, the ontology itself is a green cylinder, which indicates some kind of storage. The question is, "what is stored?" The answer is all of the details of the knowledge. However, the methodology of storage is not defined by its being an ontology.

The ontology knowledge could be stored as an English language description of all of the classes and their relations; however, this would be cumbersome and subject to misinterpretation and would not support easy computer access.

Many people have worked on the problem of storing an ontology and how to use the results. An excellent, short, and readable book by Lee Lacy describes ontologies and the Web Ontology Language (OWL) (Lacy, 2005). Briefly, given a desire to draw inferences from instances within a domain (think of the iPhone Siri's ability to

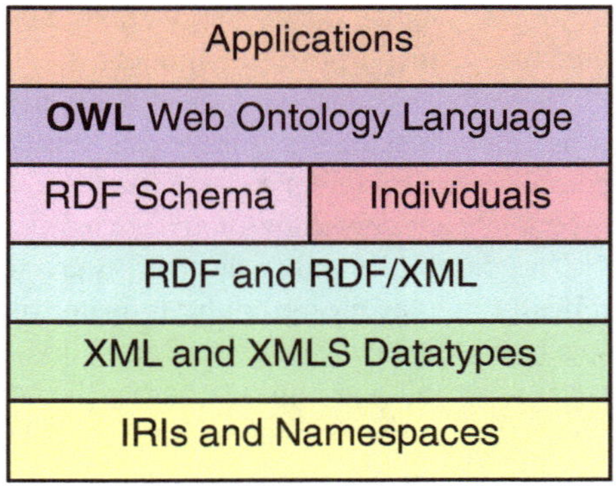

Fig. 5.6 Ontology languages

answer questions, an application similar to that in the top level of Fig. 5.6), these people developed OWL from a series of increasingly basic languages (including XML toward the bottom of the figure).

OWL has several variants; however, the basic property is that the types of knowledge that will be gathered for an ontology can be expressed in the OWL language with no misinterpretations. Further, the rules of the language permit unambiguous inferences to be drawn by computer logic. However, these same rules restrict some of the knowledge about the domain from being expressed in OWL. Because OWL is based on XML, the actual OWL files are difficult to work in the absence of supporting software. The Protégé tool is the preferred method for editing and displaying ontologies (Protégé n.d.).

OWL is not the only way to store an ontology. A relational database, such as Access, can also serve the purpose. The tables of the database will hold the data and the queries will implement the relations. With the inclusion of VBA modules, most of the desired functionality of an ontology language can be implemented. This technique requires creation of this functionality and does not supply the transparency of the OWL language; however, it does permit the storage and expression of the knowledge that is not possible in OWL.

Sources for the IW Ontology

Of all the things we could be discussing, we have reduced our attention to unconventional conflict—that is the domain of discourse—that is our "world." An ontology is a tool for capturing, retaining, and expressing our knowledge of this domain. Further, it is a tool to aid in our understanding of the domain. And, because our knowledge is always incomplete, the ontology will always be incomplete. This means that the ontology is a model—an abstraction of reality. While the ontology as a description of the domain is incomplete, the ontology as a model could be complete. That is, the ontology could include all of the parts of the domain that are germane to its purpose as a model.

In the discussions that follow, some parts describe what has been done; however, some parts discuss the ontologies (parts of the overall ontology) as if they are complete and correct, the ideal situation. In fact, it would be virtually impossible to ascertain that a given ontology is complete and correct.

The ontologies under discussion are set at a certain level of granularity. Theoretically, the level of granularity is consistent throughout the ontology. Practically speaking, this consistency has probably not been achieved. Partly, this is the case because defining granularity across the entire domain is difficult, if not impossible. Some choices are evident, for example, the cognitive processes of individuals occur at a finer level of granularity than is desired for ontologies of theater-level unconventional conflict, and these processes are not included.

The ontology that is described in this chapter is the result of two projects for the U.S. Army Training and Doctrine Command (TRADOC) Analysis Center (TRAC) and two independent research and development projects by Hartley Consulting. The two TRAC projects produced two final reports (Hartley & Lacy, 2011, 2013b) and a book chapter (Hartley & Lacy, 2013a). The results of the independent projects and simple versions of the TRAC work can be found on the Hartley Consulting web site (Hartley, 2016b).

Although Lee Lacy and I created the IW Ontology, we drew content from every source that we could find at the time.

Doing Windows Network

The Doing Windows taxonomy is described in Fig. 5.7. The variables are contained in the four influence nets (Hayes & Sands, 1997). This book was the result of three workshops of OOTW experts, looking for rational indicators of success in prosecuting Operations Other Than War. Their work provided the first 119 metrics. A listing of the Doing Windows network can be found at http://drdeanhartley.com/HartleyConsulting/IWMetricOntology/rptDoingWindowsNetwork.pdf.

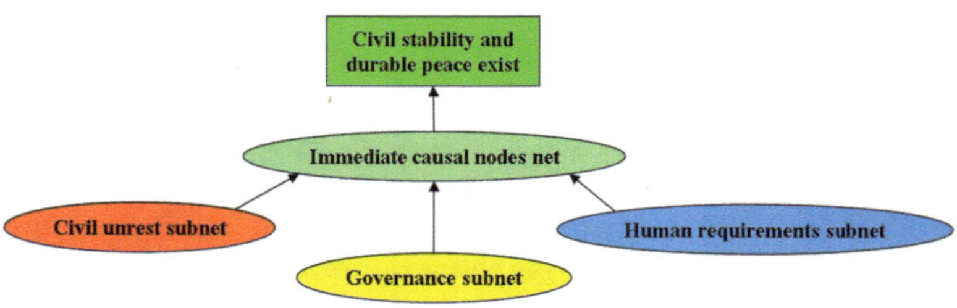

Fig. 5.7 Doing windows network design

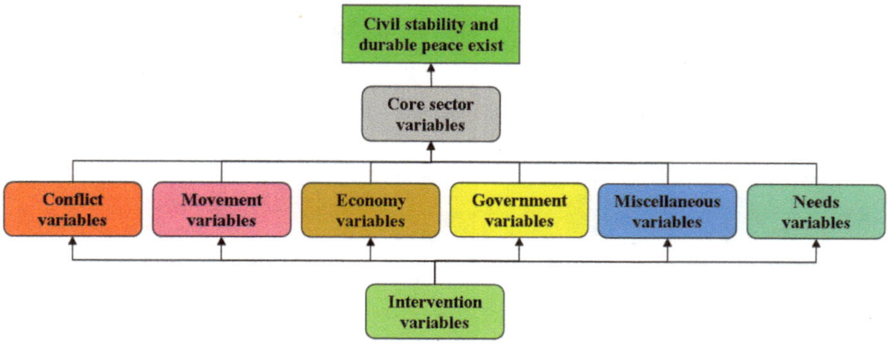

Fig. 5.8 ISSM network design

ISSM Network

The Interim Semi-static Stability Model (ISSM) taxonomy is described in Fig. 5.8. There are three levels to the taxonomy, each containing the influence variables (Hartley, 2006a). See Chapter 3, Models for a description of ISSM. The main addition to the Doing Windows list was a complete list of DIME actions (and their implied MoPs). It also added a few metrics of concern to scenarios that differ from those considered in *Doing Windows*. These DIME actions were derived from reading numerous reports of activities in OOTWs, including Bosnia, and the UN and USAID reports of activities in several actual OOTWs, including the counter-insurgency phase of the second Iraq conflict following the end of "major combat operations." This brought the number of metric type entries to 195. A listing of the ISSM network can be found at http://drdeanhartley.com/HartleyConsulting/IWMetricOntology/rptISSMNetwork.pdf.

VV&A Tool Ontology

The DIME/PMESII VV&A Tool ontology is described in Fig. 5.9. There are three levels to the ontology, with the variables connected to the lowest level shown in the diagram. DIME variables are connected both to the Status box and to the DIME box (Hartley, 2008). The tool itself is described in Chapter 6. This ontology expanded the scenario coverage and added kinetic physical environment considerations to the list. This tool also added a few items of concern to additional scenarios. The total number of entries was increased to 285. A listing of the VV&A Tool ontology can be found at http://drdeanhartley.com/HartleyConsulting/IWMetricOntology/rptVVAToolOntology.pdf.

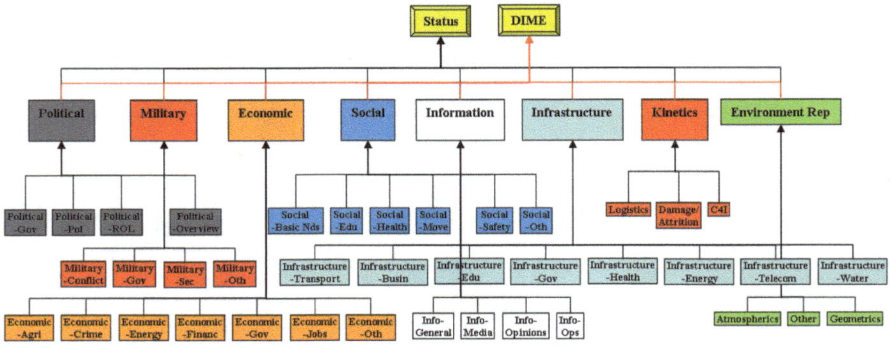

Fig. 5.9 VV&A tool ontology design

Fig. 5.10 Haskins list design

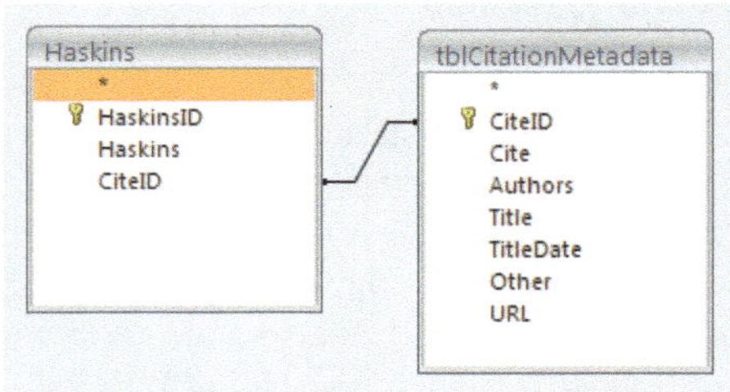

Haskins List

The list is described in the relational database design in Fig. 5.10. There is only one level to the list, with a text description (Haskins, 2010). This list produced 15 useful entries, principally of a cultural nature. A copy of the Haskins list can be found at http://drdeanhartley.com/HartleyConsulting/IWMetricOntology/rptHaskinsList.pdf.

OCRS Taxonomy

The U.S. Department of State's Office of the Coordinator for Reconstruction and Stabilization (OCRS) Essential Tasks taxonomy is described in the relational database design in Fig. 5.11. There are three levels to the taxonomy, each with a text description (OCRSLev1, OCRSLev2, and OCRSMatrixItem) (Department of State, 2005). This taxonomy of essential tasks produced 217 useful entries. A listing of the OCRS taxonomy can be found at http://drdeanhartley.com/HartleyConsulting/IWMetricOntology/rptOCRSTaxonomy.pdf.

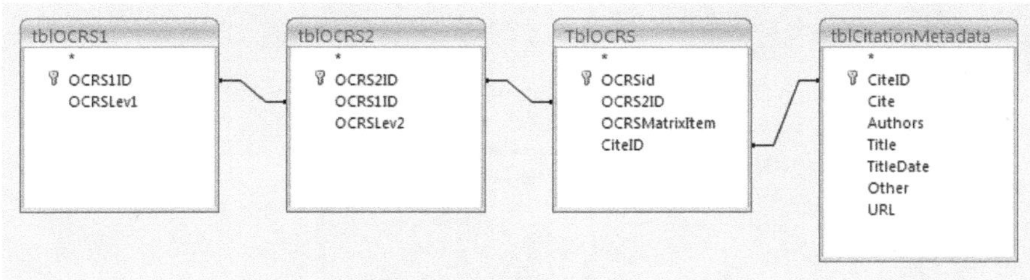

Fig. 5.11 OCRS taxonomy design

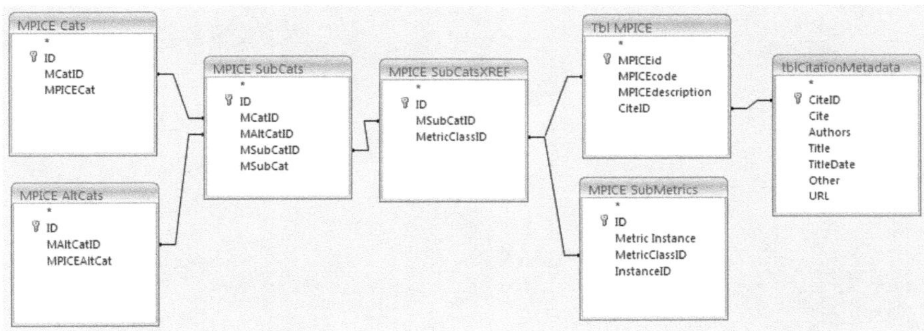

Fig. 5.12 MPICE ontology design

MPICE Ontology

The Measuring Progress in Conflict Environments (MPICE) ontology is described in Fig. 5.12. There are four levels to the ontology, each with a text description (MPICECat and MPICEAltCat at the top level, MSubCat at the second level, MPICE description at the third level, and Metric Instance at the bottom level) (Dziedzic, Sotirin, & Agoglia, 2008). This ontology of metrics produced 62 useful entries. A listing of the MPICE ontology can be found at http://drdeanhartley.com/HartleyConsulting/IWMetricOntology/rptMPICEOntology.pdf.

Hillson Taxonomy

The Hillson taxonomy is described in the relational database design in Fig. 5.13. There are three levels to the taxonomy, each with a text description (HillsonCategory, HillsonItem, and MOFE Indicator) (Hillson et al., 2009). This taxonomy produced 54 useful entries. A listing of the Hillson taxonomy is given at http://drdeanhartley.com/HartleyConsulting/IWMetricOntology/rptHillsonTaxonomy.pdf.

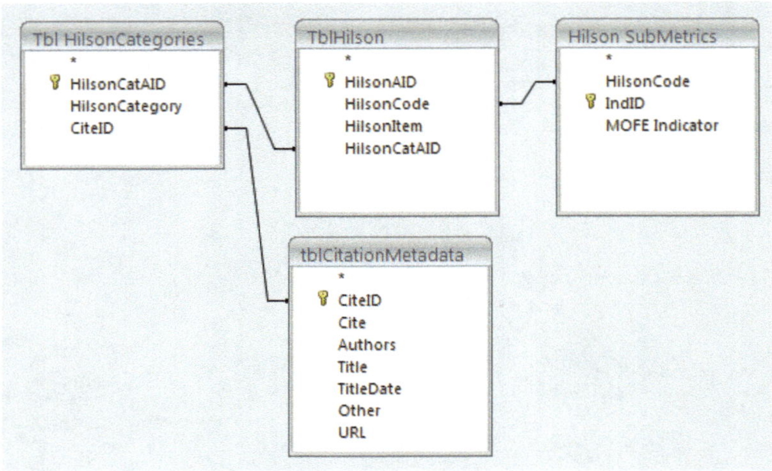

Fig. 5.13 Hillson taxonomy design

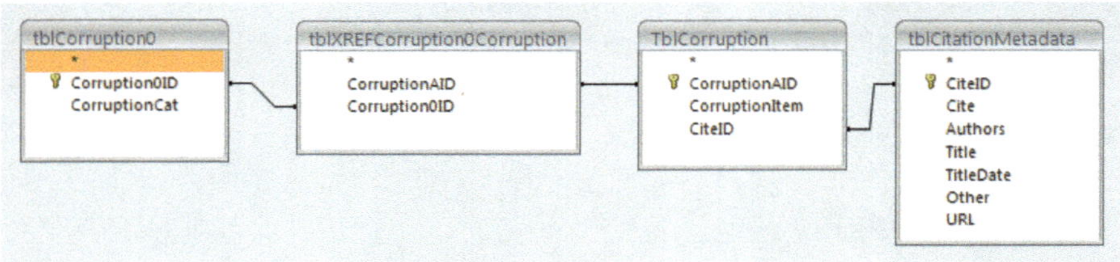

Fig. 5.14 Corruption model ontology design

Corruption Model Ontology

The ontology for the Corruption Model, developed from a conceptual model produced in a National Defense University (NDU) workshop, is described in the relational database design in Fig. 5.14. There are two levels to the ontology, each with a text description (CorruptionCat and CorruptionItem) (Hartley, 2010b). The model itself was described in Chapter 3, Technical Modeling Approaches and Fig. 3.10 shows a graphical depiction of the model. This ontology produced 131 useful entries. A listing of the corruption model ontology is given at http://drdeanhartley.com/HartleyConsulting/IWMetricOntology/rptCorruptionOntology.pdf.

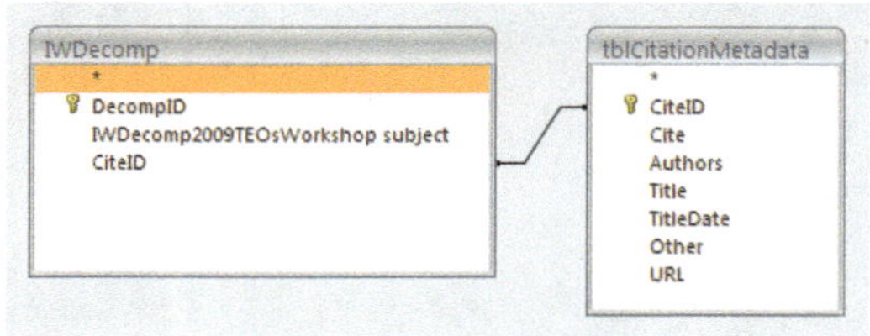

Fig. 5.15 IWDecomp list design

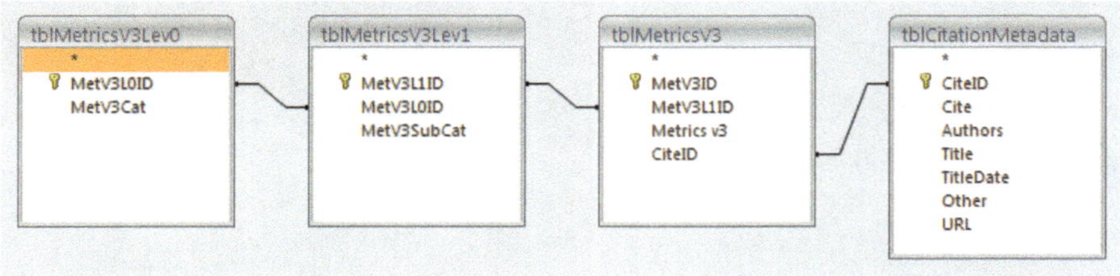

Fig. 5.16 Metrics V3 taxonomy design

IW Decomposition List

The TRAC 2009 IW Decomposition Task-Event-Outcome (TEO) list is described in the relational database design in Fig. 5.15. There is only one level to the list, with a text description (TRAC 2009, Jan). The cited work is a report on a TRAC workshop on IW. This list produced 99 useful entries. A copy of the IW Decomposition list is given at http://drdeanhartley.com/HartleyConsulting/IWMetricOntology/rptDecompList.pdf.

Metrics V3 Taxonomy

The TRAC Metrics V3 taxonomy is described in the relational database design in Fig. 5.16. There are three levels to the taxonomy, each with a text description (MetV3Cat, MetV3SubCat, and Metrics v3) (TRAC, 2010). This taxonomy was compiled by TRAC as a set of candidate metrics. It provided 226 useful entries. A listing of the Metrics v3 taxonomy is given at http://drdeanhartley.com/HartleyConsulting/IWMetricOntology/rptMetricsV3Taxonomy.pdf.

HSCB Taxonomy

Mitre's Human Social Cultural Behavior (HSCB) taxonomy is described in the relational database design in Fig. 5.17. There are five levels to the taxonomy, each

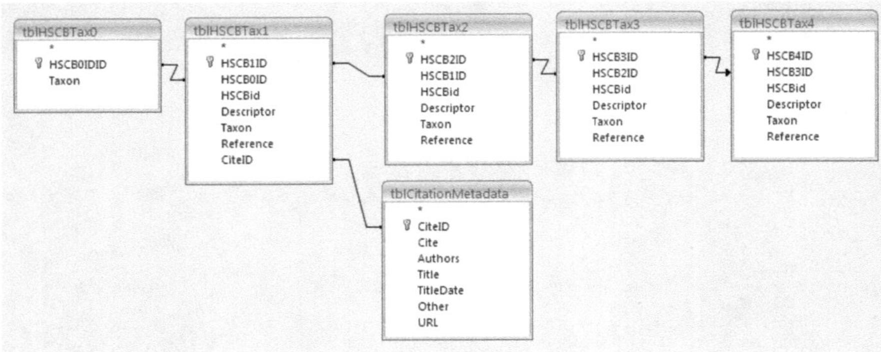

Fig. 5.17 HSCB taxonomy design

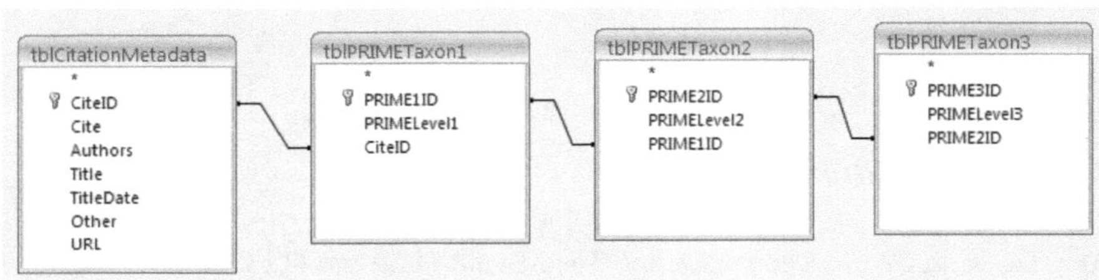

Fig. 5.18 PRIME taxonomy design

with a text description (Taxon) (Klein, 2011). This taxonomy provided 345 useful entries. A listing of the HSCB taxonomy is given at http://drdeanhartley.com/HartleyConsulting/IWMetricOntology/rptHSCBTaxonomy.pdf.

PRIME Taxonomy

SRI International's Probative Rapid Interactive Modeling Environment (PRIME) taxonomy is described in the relational database design in Fig. 5.18. There are three levels to the taxonomy, each with a text description (PRIMELevel1-3) (Lowrance & Murdock, 2009). This taxonomy provided 59 useful entries. A listing of the PRIME taxonomy is given at http://drdeanhartley.com/HartleyConsulting/IWMetricOntology/rptPRIMETaxonomy.pdf.

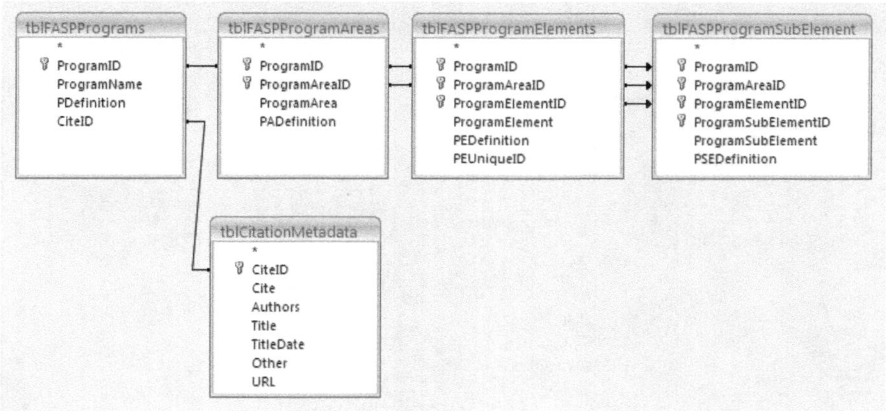

Fig. 5.19 FASP taxonomy design

FASP Taxonomy

The Department of State and USAID Foreign Assistance Standardized Program (FASP) Taxonomy is described in the relational database design in Fig. 5.19. There are four levels to the taxonomy, each with a text description (ProgramName, ProgramArea, ProgramElement, and ProgramSubElement). A listing of the FASP taxonomy can be found at http://drdeanhartley.com/HartleyConsulting/IWMetricOntology/rptFASPTaxonomy.pdf. The 2006 version of the FASP document was the most current when the ontology was being created. It may be found at http://drdeanhartley.com/HartleyConsulting/IWMetricOntology/FAF%20SPSDef-2007.pdf. The current version is the 2016 update (Department of State, 2016).

Situation-Independent Ontology

The central organizational principle for the development of the IW Ontology is provided by the context diagram in Fig. 5.20. The Operational Environment that includes everything relevant to unconventional conflict is divided into three parts: Actors, Actions, and the Environment.

Actors are human and natural entities that cause things to happen, thereby changing things. Actions are the interventions, events, and ongoing processes that are performed by actors, which directly cause changes. The Environment represents the passive entities in the Operational Environment.

Actors perform Actions, which affect the Operational Environment (OE). The state of OE, including any changes, is described by State Variables (also called metrics). Actors perceive the OE by means of the State Variables. State Variables

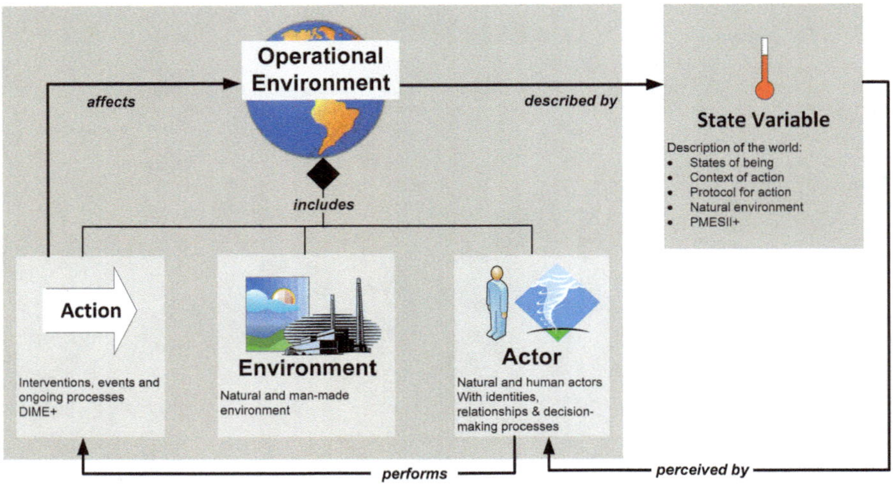

Fig. 5.20 Context diagram

include both numeric variables (true metrics) and categorical variables (e.g., type of government).

The divisions presented in Fig. 5.20 provide a starting place for building and understanding the ontology. We can construct subontologies for actors, actions, environment, and metrics. As a matter of terminology, we will call the lowest level classes of the actor, action, and environment ontologies "elements." We will call the lowest level classes of the metric ontology "metrics." All of the elements and metrics are derived from the sources, which were described earlier. They are linked to the appropriate sources in order to improve the understanding of each class.

Metric Ontology

The metric ontology is based directly on the PMESII+ paradigm. This paradigm is meant to support a description of the state of a situation, which is precisely what the state variables (metrics) do in detail. Thus, each metric is related to a PMESII subcategory by the "is-a" relation.

Figure 5.21 lists the 8 categories of PMESII+ and the 37 subcategories and indicates that there are 750 metric classes at the next level. Table 4.2 on page 105 gives some example metrics.

Figure 5.22 provides a sketch of the metric ontology, repeating the items discussed earlier, but adding connections and color coding. All of the metrics are represented by a single class with a vertical arrow representing the connection(s) of the metric to the PMESII subcategories and with a horizontal arrow representing the connections to other things in other ontologies. The sketch appears in the form of a

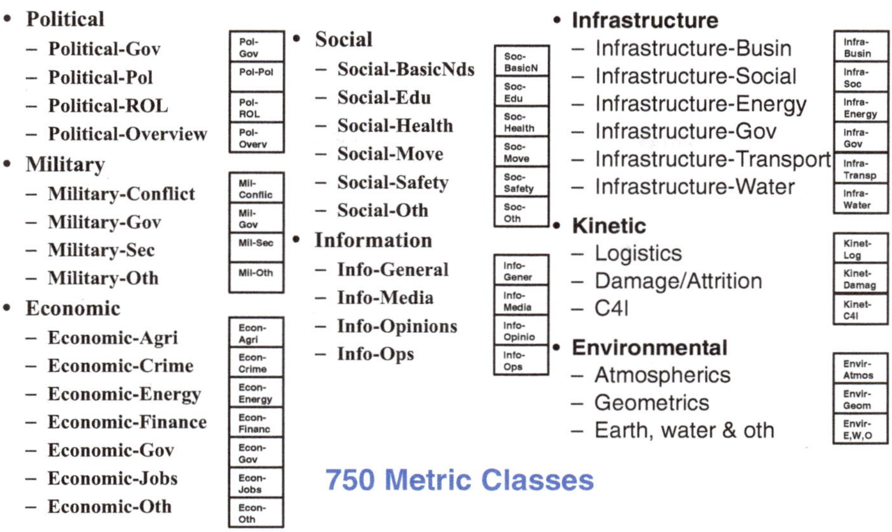

Fig. 5.21 Listing of the metric categories and subcategories

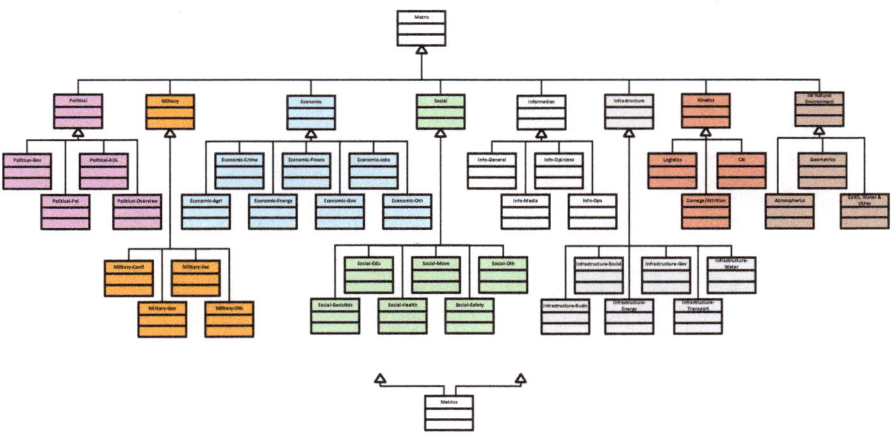

Fig. 5.22 Metric ontology

taxonomy; however, it does not show the connections of some metrics to multiple parents.

To download a file containing the metric ontology, go to http://drdeanhartley.com/HartleyConsulting/IWMetricOntology/IWOntology2.htm and click on the download link in the description to the left of the metric ontology figure.

- **Conceptual**
  - **GeoPoliticalDivision**
  - **EconomicDivision**
- **Individual**
  - **OtherIndividual**
  - **KeyLeader**
- **Significant Group**
  - **SocialOrganization**
  - **EconomicOrganization**
  - **ArmedForce**
  - **PoliticalOrganization**
- **Demographic Group**
  - **StaticPopulation**
  - **MobilePopulation**
- **Physical**
  - **Vehicle**
  - **Environmental**

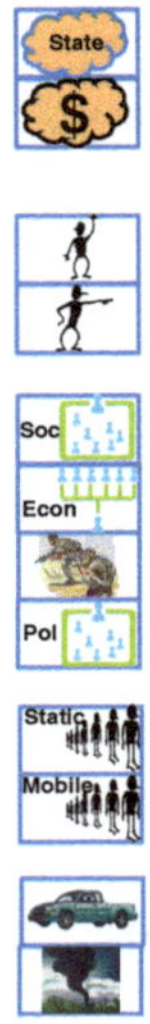

102 Actor Classes

Fig. 5.23 Listing of the actor categories and subcategories

Actor Ontology

Figure 5.23 lists the 5 actor categories and the 12 actor subcategories. Each of the 102 actor elements is connected to an actor subcategory by an "is-a" relation. Note that at this point, there is no connection to the PMESII paradigm: the categories and subcategories describe types of actor elements. For example, the distinction among individuals, significant groups, and demographic groups, described in Chapter 2, is made at the category level. Within the individual category, the subcategories distinguish between key leaders and other individuals. There are different actor classes for key leaders of various types of groups. The "other individual" subcategory exists to allow for modeling "bit players" who are temporarily important, such as a suicide bomber.

Figure 5.24 provides a sketch of the actor ontology as a taxonomy, repeating the aforementioned items, but adding connections and color coding. The single class below the taxonomy stands for all of the actor classes.

To download a file containing the actor ontology, go to http://drdeanhartley.com/HartleyConsulting/IWMetricOntology/IWOntology2.htm and click on the download link in the description to the left of the actor ontology figure.

Action Ontology

Figure 5.25 lists the 7 action categories and 17 subcategories and indicates there are 385 action elements. These action elements constitute the classes that would be the lowest level of a DIME ontology.

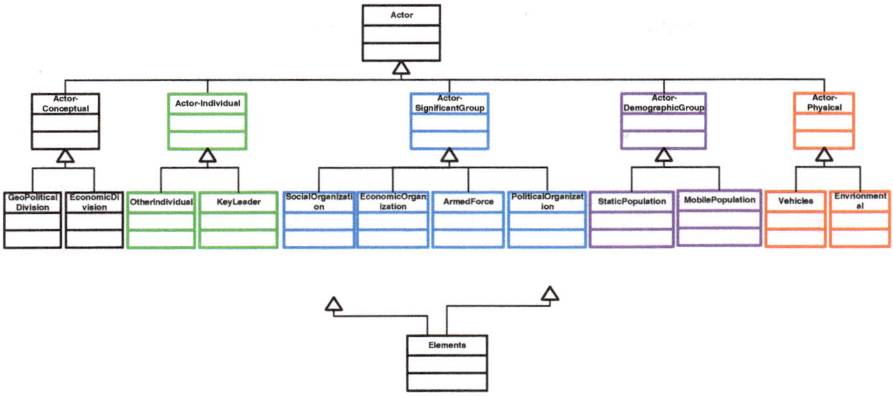

Fig. 5.24 Actor ontology

Fig. 5.25 Listing of the action categories and subcategories

Unfortunately, the connection between the set of action classes and the DIME paradigm is fuzzy, at best. A few actions may be clearly Diplomatic, Informational, Military, or Economic. A few more may be clearly connected to more than one of these, a situation which is easily handled in an ontology. However, most actions have poorly defined connections. For example, a negotiation action may be handled by a civilian diplomat, a military unit, or a civilian economic organization, depending on the situation. Further, some actions may be initiated in one D-I-M-E category and carried out in another category.

The concept of DIME (or any of its variants) as indicating that there are different levers of governmental power is useful. However, decomposing DIME down to the

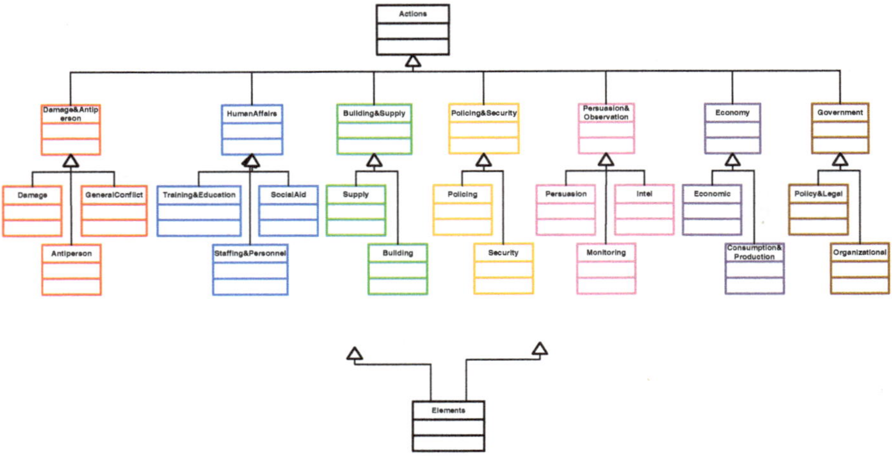

Fig. 5.26 Action ontology

actions level is essentially meaningless. Creating connections that link almost everything to each of the DIME categories does not clarify the understanding of what is going on.

Figure 5.26 provides a sketch of the action ontology as a taxonomy, repeating the aforementioned items, but adding connections. The single class below the taxonomy stands for all of the action classes.

To download a file containing the action ontology, go to http://drdeanhartley.com/HartleyConsulting/IWMetricOntology/IWOntology2.htm and click on the download link in the description to the left of the action ontology figure.

Environment Ontology

Figure 5.27 lists the 4 passive environment categories and 21 subcategories. The 209 environment classes at the next level provide the first example among the elements of the need for an ontology, as opposed to a taxonomy. One of the natural environment classes is wild fire, which can be natural or manmade, requiring two parents at the subcategory level.

Figure 5.28 provides a sketch of the environment ontology as an ontology, repeating the items aforementioned, but adding connections and color coding. The single class below the taxonomy stands for all of the environment classes.

Once the three element ontologies are combined, a second example of the need for a more general ontology, rather than a taxonomy, becomes apparent: some of the elements appear in more than one of the element ontologies. For example, depending on the model, a tornado might be modeled as an actor, an action, or merely part of the environment.

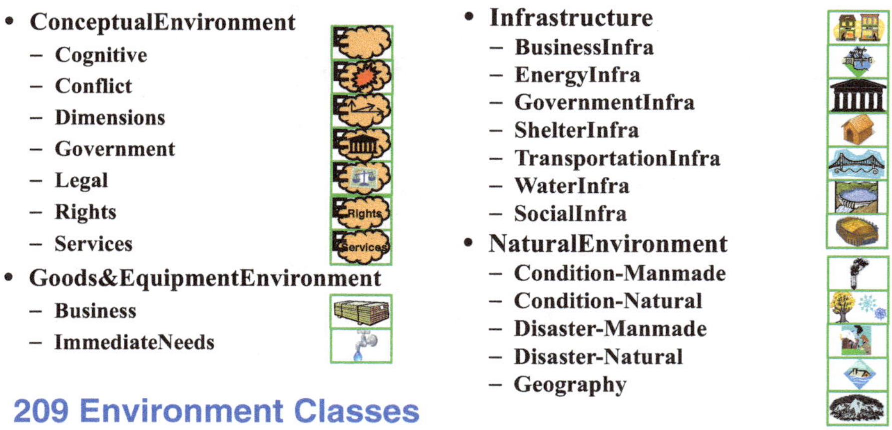

Fig. 5.27 Listing of environment categories and subcategories

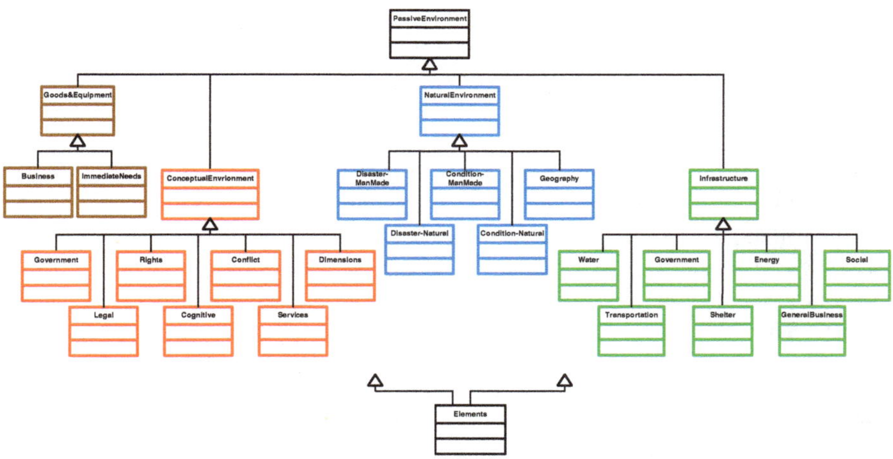

Fig. 5.28 Environment ontology

To download a file containing the environment ontology, go to http://drdeanhartley.com/HartleyConsulting/IWMetricOntology/IWOntology2.htm and click on the download link in the description to the left of the environment ontology figure.

Connecting Elements to Metrics

The context diagram, Fig. 5.20, shows the operational environment being described by the state variables—the metrics. These connections are not made at the element

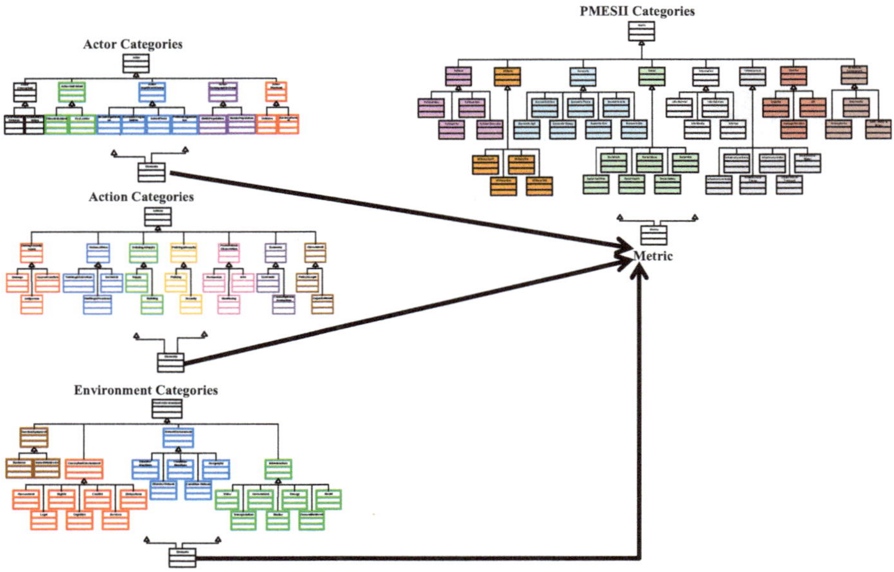

Fig. 5.29 Connecting elements to metrics

category or subcategory levels, but at the element class level—to the metric classes, as appropriate. This level of connection is illustrated in Fig. 5.29.

The actor and environment elements are connected to metrics ending in "rating," as shown in Table 4.2 (in Chapter 4, page XX). These metrics describe these elements' state of being. Action elements are connected to metrics ending in "MoP". The last entry in Table 4.2 is connected to an action element pertaining to providing external government advisors. Measures of performance describe the immediate results of an action, as opposed to a measure of effectiveness that would describe the effectiveness of the action. All elements are connected to at least one metric and some are connected to more than one metric. These connections are neither "is-a" nor "part-of" relations. They are "has-metric" relations.

A little reflection will show that this connection exposes a profound point. Suppose we were observing a real situation. The values in the set of all metrics, such as the *CentralAuthorityEffectivenessRating* metric (Table 4.2), would describe the current state of affairs (assuming we had some way of knowing the values). The values at a later time would describe the state at that time and the differences in the values would describe the change. The rating of the effectiveness of the central authority does not seem particularly remarkable when looked at this way. However, now consider that we are getting these values from a simulation. The rating value does not change by magic, but rather because of some theory or theories that declare that it should change in a particular way because of the change of some set of other variables. This is theory-based connection.

This is a key part of the entire ontology paradigm. If the metrics tell us the situation has changed, the theories that describe the causes of these changes are the theories of our model. That means the each metric represents an attachment point of one or more theories, ***which are outside of the ontology***.

To download a file listing the metrics connected to each element, go to http://drdeanhartley.com/HartleyConsulting/IWMetricOntology/IWOntology2.htm and click on the download link in the description to the left of the connecting elements to metrics figure.

Properties

Properties are the variables of a class. An instance of a class is an object and its corresponding variables are called attributes. Each of the element classes has a property named *entityIdentity*. Having this property in common is part of what defines this class. Each instantiation (object) has a different value of its *entityIdentity* attribute—its name. Classes and objects also have behaviors or operations. For example, some classes represent things that can move—change their location (another property). For simplicity in this discussion, operations will be lumped in with properties.

Subclasses of a class inherit the properties of a class and add additional properties that help define the subclass. Figure 5.30 illustrates this inheritance process by showing the properties of all elements (including the existence of a link to one or more metrics) and showing additional properties of the subclasses down the chains.

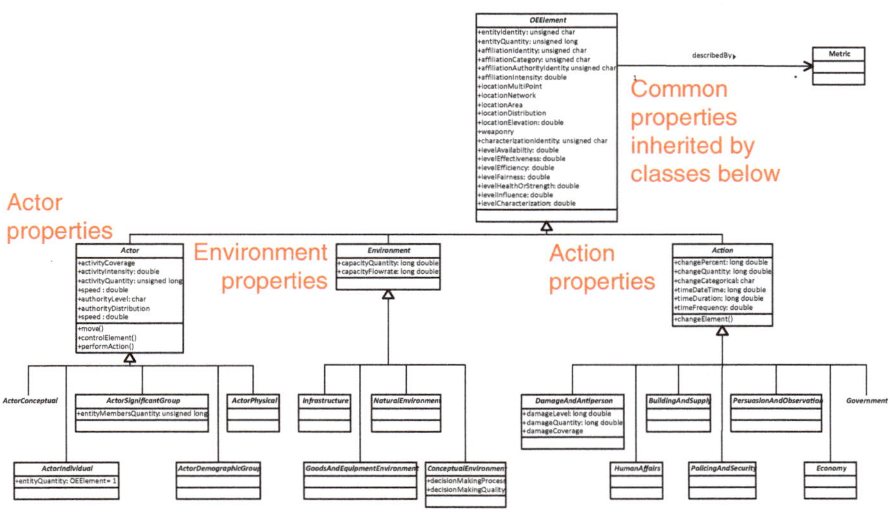

Fig. 5.30 Inheritance of properties by elements

(You are not meant to be able to read the individual properties in this figure. They are listed and described in Chapter 7, Gathering the Data.)

The properties are best defined in terms of the elements. For example, a vehicle class element clearly must have the property of being capable of movement. These properties convey state information that is not directly connected to the PMESII ontology as are the metrics.

Some properties represent fixed values for their instances—an object will not change its name; however, some represent current values—the location of an object may vary over time. We will revisit this topic in the data section in Chapter 7.

To download a file containing the property ontology, go to http://drdeanhartley.com/HartleyConsulting/IWMetricOntology/IWOntology2.htm and click on the download link in the description to the left of the property ontology figure.

Stocks and Flows

Recall that the purpose of an ontology is to describe what we know about a domain. In this chapter, we are describing the structure of the ontology, not the contents of the ontology. However, the structure is driven by the contents. We wish to have a structure that allows us to include everything we know about unconventional conflict that is relevant to modeling it.

Upon investigating the contents of the ontology so far, the various elements and metrics, we find that there are connections among these classes that are evident because we know what the classes refer to, but which are not yet expressed in the ontology. One set of these connections has a counterpart in system dynamics—stocks and flows. For example, there is an environment element for bridges and tunnels that has a capacity property. There are also two actions that relate to this, destroy bridges and tunnels and build bridges and tunnels. These two actions reduce capacity and increase capacity, respectively. These elements are related in the real world, and we need to introduce this relation into the ontology.

Figure 5.31 illustrates the structure of this type of stocks-and-flows relation. One or more actions are related to an environment element as increasing the capacity of the environment element and one or more actions are related as decreasing the capacity. Not all actions and not all environment elements are found in the set of all environmental stocks-and-flows relations.

We also find a similar, but more complex, stocks-and-flows relationship that relates to organizations (Fig. 5.32). The first part is very similar, with an actor element that is an organization taking the place of the environmental element. The

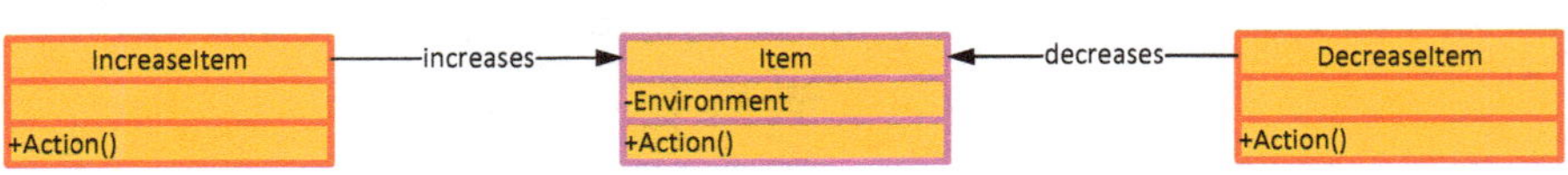

Fig. 5.31 Environmental stocks-and-flows relations

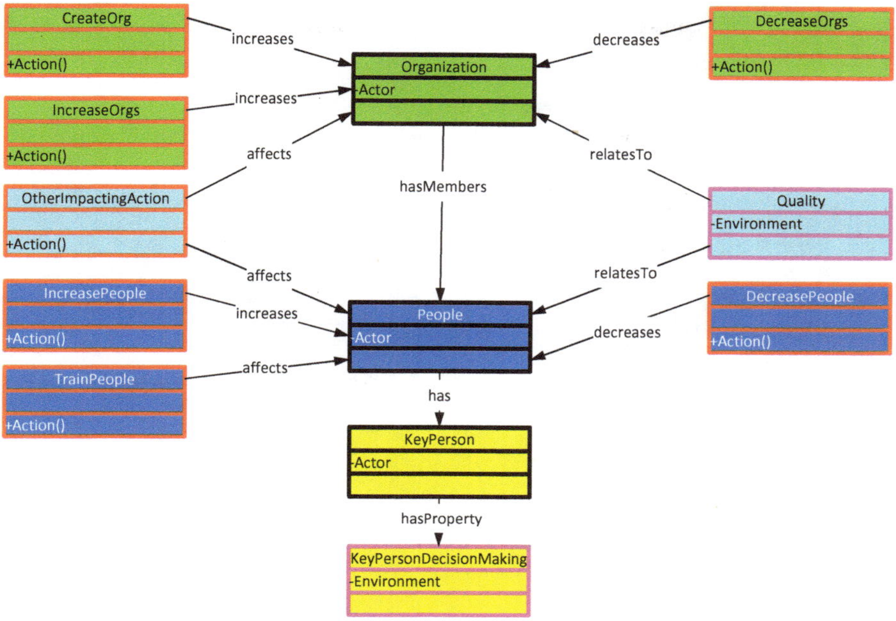

Fig. 5.32 Organizational stocks-and-flows relations

increase and decrease of the organization refer to increasing or decreasing the structural size of the organization (more or fewer military units, more or fewer businesses of a given type, etc.). The additional action of creating the organization is a special case of increasing it.

However, organizations are made up of people, actors that are different from organizations, so that an additional actor class has to be linked with a hasMembers relation. The number of people can be increased or decreased (additional actions). In some special cases the people can be trained (actions that are included in the ontology for certain organizations, such as military units, but not for all organizations, such as private businesses).

In addition, there are other actions that impact some organizations, which need to be linked. Also, some organizations have environmental elements that describe the quality of the organization or its people. These elements are linked, as appropriate.

Finally, many significant organizations have one or more keyperson actors that need to be linked and these keypersons have environmental elements reflecting their decision making.

The nature of the stocks-and-flows ontology can be illustrated by the following partial list of the stocks-and-flows sets in the ontology:

- Organizational Oriented
  - Intervention

    InterventionForce: an intervention force
    InterventionOrg: an intervention organization (e.g., composed of diplomats, etc.)
    I*nterventionSupport*
  - HostNation

    HNExecutive
    HNLegislative
    HNBureaucracy
    HNMilitary
    HNLawEnforcement
    ...
  - OtherForces

    Armed_RegimeSponsoredNonMilitary
    ...
  - Economic

    Business_Agriculture
    ...
  - OtherPeople

    NGOOrg
    ...
- Environmental Oriented
  - Infrastructure

    Transport_Airport
    ...
  - OtherItems

    Crime_Corruption
    ...

As this list shows, there are two types of stocks and flows and six categories, split four and two. The 94 stocks-and-flows subcategories are split 47 organization oriented and 47 environment oriented. Because these are classes, they can be instantiated as multiple objects, if needed. Altogether, these stocks-and-flows links reflect knowledge about the situation that can now be incorporated into the ontology.

To download a file containing the stocks-and-flows ontology, go to http://drdeanhartley.com/HartleyConsulting/IWMetricOntology/IWOntology2.htm and click on the download link in the description to the left of the two stocks-and-flows figures.

Semantic Thesaurus

There is one more type of knowledge that is pertinent to the situation that we need to incorporate into the structure of the ontology. This knowledge is obvious to a human examining the metrics, but not to a computer. For example, a human knows that roads, bridges and tunnels, and rail lines are all related. These connections could have been made among the element classes; however, it developed that it was more convenient to make the connections among the metrics and use the links between metrics and elements to create the connections among the elements as indirect connections. Figure 5.33 shows that each metric is linked to several semantic terms and Fig. 5.34 shows that each semantic term is linked to several metrics. Together these linkages result in a semantic thesaurus connecting metrics to other metrics with similar meanings and, ultimately, elements to other elements.

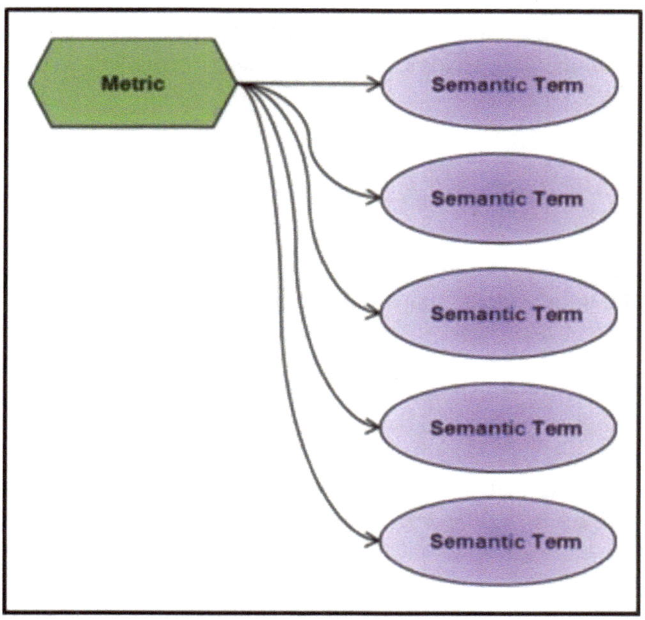

Fig. 5.33 Metric to semantic term

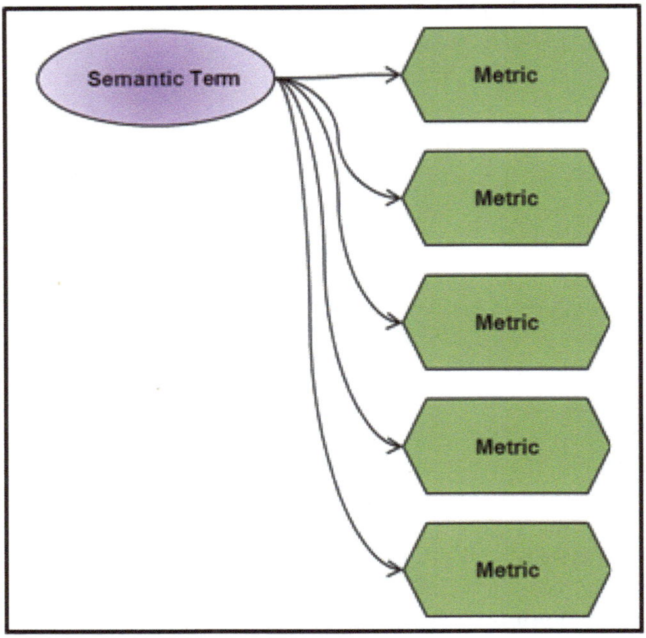

Fig. 5.34 Semantic term to metric

To download a file containing the semantic terms, go to http://drdeanhartley.com/HartleyConsulting/IWMetricOntology/IWOntology2.htm and click on the download link in the description to the left of the semantic thesaurus figures.

Recap of What We Can Say About What We Know

The structures we have defined up to now allow us to capture a large amount of knowledge about unconventional conflict. Figure 5.35 illustrates some of this.

Following the list of instructions in the upper right of the figure will help to understand it. First, select a metric, here the large orange octagon, labeled "*BridgeAndTunnelCapacityRebuiltMoP*," and define it, which means link it to its definition (the paper shape) and to the source(s) from which it comes (the yellow rectangles). Second, make the PMESII links (vertical arrows below the octagon, leading to Economic and Infrastructure shapes at the bottom). Third, link to the element (the rounded rectangle representing the action "Rebuild bridges and tunnels"). Fourth, define the element, which means link it to its definition (the paper shape) and link it to its sources (the yellow rectangles). Fifth, make the PMESII links (by connecting to all metrics connected to this element [not shown in the figure] and showing their PMESII links, magenta arrows). Sixth, make the links to

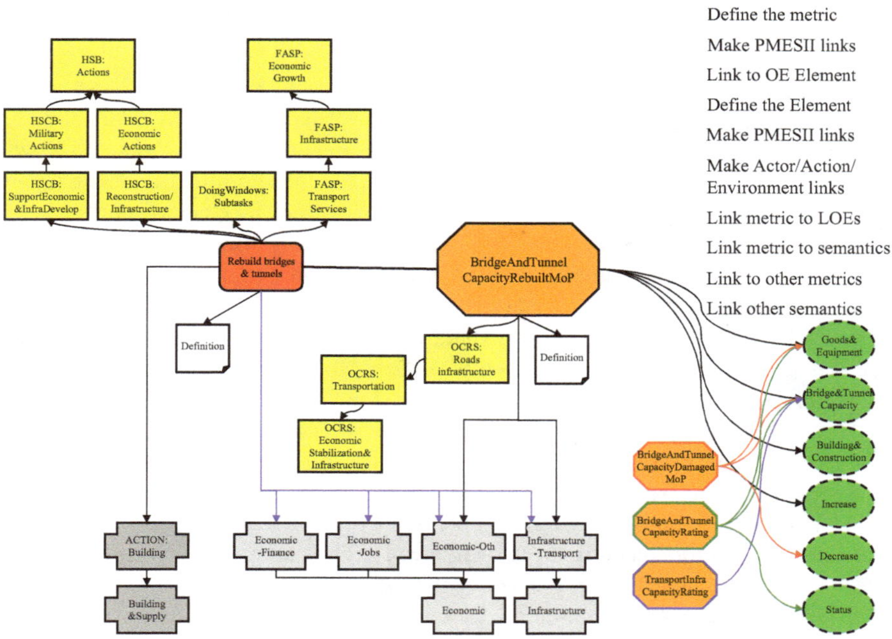

Fig. 5.35 Example ontology links

the action ontology categories (shapes at the lower left). Seventh, link the original metric to its semantic terms (first four green ovals). Eighth, link to other metrics (the smaller orange octagons). Ninth, link to other semantics (remaining green ovals).

The point is that the knowledge contained in the ontology covers a wide range of information about the domain, and this knowledge can be recovered by following the relationships.

Situation-Dependent Ontology

Despite the large amount of knowledge that is encodeable in the ontology structure to this point, it is not enough. The structure to this point permits the description of situation-independent knowledge. For example, bridges and tunnels are related to roads no matter what the situation is. However, there are things that depend on the situation, which require special structures.

GTO Structures

The number of parties or sides that are relevant to a particular situation is situation dependent; however, their existence and certain information about them are already covered at the instantiated object level. For example, the stocks-and-flows *InterventionForce* class allows for multiple countries' armed forces and associated information. However, each major player in the situation (labeled "owner") has its own agenda, consisting of goals and tasks that the owner thinks will accomplish these goals. At the lowest levels, the goals will decompose to metric classes and the tasks will decompose to action classes while the owner corresponds to an actor class. A Goal-Task-Owner (GTO) structure can be defined without knowledge of the situation—and will be defined in this section. However, GTO structures cannot be populated without situation-specific knowledge.

Figure 5.36 illustrates a typical scenario or situation. The action takes place within a host nation, with its geography, resources, and populace—and its culture, religions, laws, etc. This arena is represented by the ellipse in the center.

This scenario or situation is an example of unconventional conflict because there are competing actors, each with its own agenda. The nature of this competition is what makes the situation neither conventional war nor peace. In this figure, the U.S. armed forces are represented as the Coalition. Other U.S. governmental agencies are represented by the State Department. Ideally, the goals of these two,

Fig. 5.36 Twelve sides to the conflict

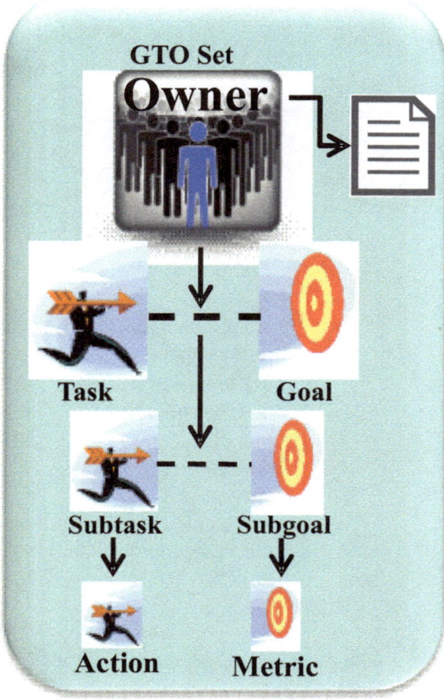

Fig. 5.37 GTO set structure

while differing in detail, are in consonance. The tasks, however, definitely differ because the capabilities of these actors differ.

The host nation (HN) is also represented by two actors, the HN government and the HN armed forces. The goals and tasks of these two actors may differ significantly. In the typical situation, there are non-governmental organizations that have significant impacts. Here an international NGO represents a fairly benign, apolitical actor and a political NGO represents an actor that sides with one or more of the opposition forces.

Also, typically, there are commercial interests in the arena. Here a construction company represents contractors hired to implement some of the tasks of the Coalition and State Department. The acquisition company represents external interests that wish to purchase HN resources. The regional power represents some external country in a manner analogous to the State Department, but with possibly opposing interests.

The three opposing forces that are represented here consist of groups with interests in opposition to the HN government. Their goals differ among themselves; however, many of their actions may be related so that they may form temporary alliances.

The quote from Sun-Tzu is apt, "Know the enemy and know yourself; in a hundred battles you will never be in peril" (Sun-Tzu, 1963). The GTO set structure, illustrated in Fig. 5.37, formalizes the concept of encoding these competing interests.

Table 5.1 Example GTO set (Part 1)

		Element ID	Element ID
GTO task	GTO subtasks	Actions—direct (TT)	Actions—indirect (ttt)
1.0 Establish Civil Security (combat ops)	1. Restore and maintain order	167, 184, 126, 182, 2054	33, 49, 51, 53, 132, 134, 136, 151, 1864, 1868
	2. Conduct operations to halt violence	126, 167, 184, 199, 1837, 1839, 2049, 2050	201, 203, 205, 209, 211, 213
	3. Establish border security	138	1822
	4. Provide public safety support	174, 211	165, 170, 172, 176, 415, 419
	5. Provide civil defense support	17, 155, 167, 2025	
	6. Perform HN police training and support	76, 80	78, 94
	7. Execute civil security common or general tasks	178, 180, 182, 184, 186, 1826, 1827	128, 130, 215, 217, 235, 327, 331, 363, 377, 445, 476, 478, 480, 482, 484, 505, 511, 517, 524, 533, 540, 564, 566, 568, 570, 584, 586, 592, 647, 649, 651, 653, 655, 657, 659, 677, 679, 681, 699, 701, 703, 705, 707, 709, 711, 714, 716, 718, 720, 722, 724, 726, 728, 730, 732, 740, 743, 745, 747, 1811, 1812, 1813, 1815, 1905, 1918, 2021, 2022

The situation or scenario is defined as a model, with a name and a date. The model has a several *GTOSetOwners* (the competing actors). Each *GTOSetOwner* is identified as an actor within the IW ontology and has a citation that includes the defining metadata for the GTO set.

The *GTOSetOwner* owns several *GTOTaskGoalPairs*, consisting of a *Goal* and a *Task* for accomplishing the goal. Each *GTOTaskGoalPair* has several *GTOSubTaskSubGoalPairs*, consisting of *SubGoals* of the *Goal* and *SubTasks* of the *Task*. Each *SubGoal* has one or more Metrics within the IW ontology and each *SubTask* has several Actions within the IW ontology. Together, the GTO Sets in a single model represent a scenario or situation.

Table 5.1 shows one task from one GTO set and Table 5.2 shows the corresponding goal. The owner (Coalition Forces) is not shown. The GTO task and its decomposition into GTO subtasks are standard military forces tasks. Note that the seventh subtask basically says, "And all other things that fit this GTO Task." Each subtask is decomposed into Action classes that are required and sufficient to accomplish the subtask. These are shown by their element ID for

Table 5.2 Example GTO set (Part 2)

	Metric ID	Metric ID		Metric ID	
GTO subgoals	Indicators (ind)	Metrics (gg)	GTO goal	Metrics (G)	Description
1. Civil unrest quelled		120	1.0 safe, secure, and stable environment established	418	Combat Operations: restore and maintain order; establish border security; provide public safety and civil defense support; train and support police
2. Violence halted	121, 122, 123	1006			
3. Borders are secure		808			
4. Public safety established		5, 173			
5. Civil defense mechanisms in place		1056			
6. HN police trained and established		71			
7. Safe, secure, and stable environment established	472, 473, 474	418			

brevity and are divided into two types, those that directly lead to accomplishing the subtask and those that are indirect or implied requirements.

The GTO subgoals that correspond to the subtasks are shown in the first column of Table 5.2. The metrics that most closely show the status of the subgoal are in the third column. The second column shows the metrics that are useful indicators of that status. All are represented by their metric ID for brevity. The last subgoal (column 1), which corresponds to the "all else" subtask, is set to equal the GTO goal, which is repeated in column 4 with its metric in column 5. Column 6 displays the verbal description of the task and goal. In the general case without a predefined decomposition, this verbal description is used to decompose the task and goal into subtasks, subgoals, actions, and metrics.

Metric Models

Each owner has a set of goals and subgoals in his agenda and a set of tasks and subtasks that he believes will lead to the realization of this agenda. Even though the owner's beliefs do not ensure that accomplishing these tasks will, in fact, lead to

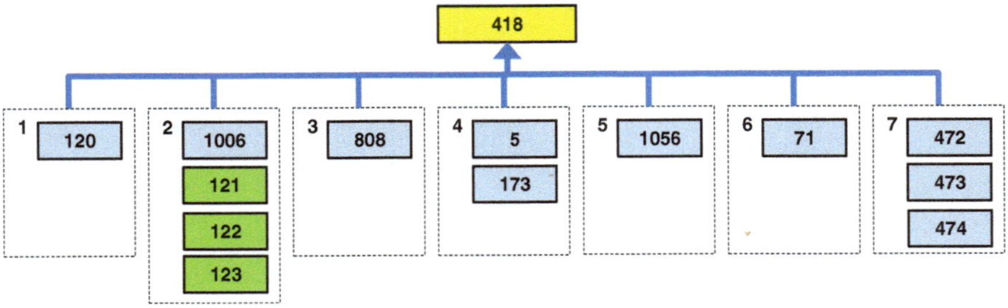

Fig. 5.38 Part of the metric model for a GTO set

accomplishing the goals, it is important to record in the ontology what each owner believes. These beliefs form the owner's metric model.

The first part of the metric model consists of the connections between the subtasks and subgoals. That is, the owner believes, for example, that "Establishing civil security (combat ops)" by means of the listed actions will result in "Civil unrest is quelled," measured by the listed metric (subtask 1 in Table 5.1 and subgoal 1 in Table 5.2). The second part of the metric model consists of the connections between the subtasks and the tasks and between the tasks and the overall agenda or, alternatively expressed, between the subgoals and the goals and between the goals and the overall agenda. Viewed in this latter fashion, the set of metrics on the goals side of the GTO set form a model that describes the owner's beliefs. As shown in Fig. 5.38, the owner believes that the value of metric 418 can be inferred from the values of the metrics below it. It should be noted that these connections are not being represented as justified by any theories; rather, they represent the beliefs of the owner. Each owner will have his own metric model and the collection of these metric models may be in disagreement.

Owner Rules

The owner in each GTO imposes a set of rules on its agents in executing the GTO set tasks. These rules are called rules of engagement in military organizations, but have other names in other organizations. These rules can be constant throughout the time span of a situation and consistent across the entire theater; however, neither is required. Figure 5.39 represents these rules symbolically.

Table 5.3 shows a sample of owner rules that could be assigned at various times and places by an owner. They are organized here by action subcategory, although other organizational principles are possible.

Note that the definitions of which actors are friends, foes, neutral, or shades of these relationships are set in the Actor Relations Structures, as are such things as ownership of or control of geographic areas, both discussed in a later section. Such concepts may be tested in the rule definitions.

Fig. 5.39 Owner rules

Table 5.3 Sample owner rules

ActionSubCategory	Rule
Antiperson	Avoid casualties to civilians, friendly forces; no chemical, biological, or radiological weapons
Antiperson	Casualties are only restricted by resources and capabilities
Antiperson	Do not injure anyone
Antiperson	Engage in limited extrajudicial, criminal, and suppression activities
Antiperson	Use weapons only when attacked; avoid civilian casualties
Damage	Avoid damage to civilian infrastructure, friendly force infrastructure, host nation infrastructure
Damage	Damage is only restricted by resources and capabilities
Damage	Do not damage the environment
GeneralConflict	Avoid all conflict
GeneralConflict	Conduct conflict by any means available
GeneralConflict	Do not participate in terrorism or piracy

AAR Structures

The Actor-Action-Result (AAR) sets represent a finer level of detail than that of the GTO Sets. Where the GTO Sets represent a scenario, an AAR set represents a

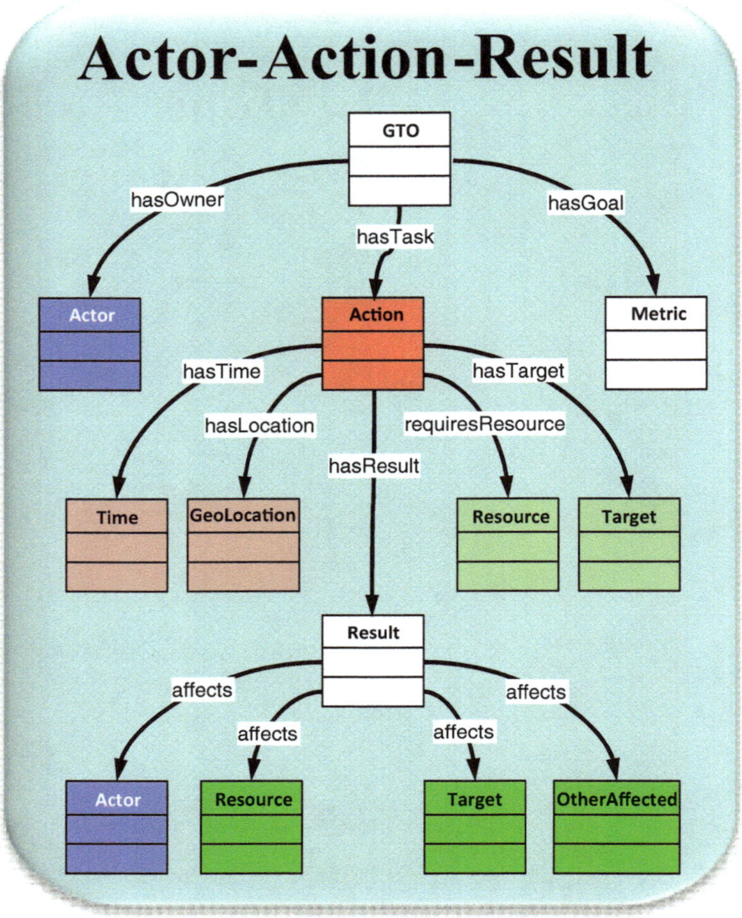

Fig. 5.40 Actor-action-result set structure

vignette within the scenario. An Action from one GTO set (or perhaps two or three very closely related Actions) and the associated Metric (or Metrics) are instantiated and form the basis for an AAR set. For example, if the Action is attack Bridges and Tunnels, the instantiated action could be to attack a particular bridge.

Figure 5.40 shows that the action (red box in the middle of the diagram) has several associated elements: time, geo-location, target, and resources needed for the action. These resources might be physical elements (vehicles, etc.) and might include other actors.

The action has a result; it affects the initiating actor, the resources, the target, and possibly other elements (bystanders, other infrastructure, etc.) and changes the values of their metrics. The rationale for computing the effects is not included.

Rather, the Result placeholder acts as a call for one or more social or physical theories that provide this rationale.

This is a key part of the entire ontology paradigm. The rest of the ontology structures and contents, including the instantiation for a specific situation, is *descriptive* modeling. That is, it involves including a description of what is known and observed about unconventional conflict in general and about a particular conflict. This structure specifically identifies the points where the model must include *causal* modeling or substitute some *empirical substitute*, ***which is not contained in the ontology***. Some of the causal modeling will be relatively straightforward, for example, the effects of a certain amount of explosive force on a bridge. However, some of the causal modeling will require social theories that are not so straightforward, for example, the changes of opinions and resulting actions that will occur among the affected people. While errors in the descriptive modeling will affect model validity, identifying and correcting such errors or mitigating their effects is much less difficult than for errors in the causal modeling. The reason is that we know much less about the workings of the real world when it comes to the causes of social effects. However, this ontological structure allows for the specific identification of where such errors may occur in the model.

Actor Relations Structures

The Actor Relation Sets (Fig. 5.41) provide the final piece of the description of the scenario or situation. The Actor-Actor and Actor-Environment structures provide for the definition of the relationships (boss/employee, leader/follower, tribe/

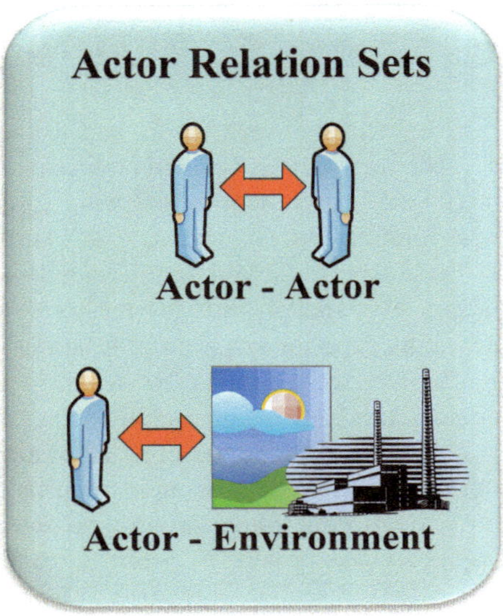

Fig. 5.41 Actor relation sets

Table 5.4 Sample actor-to-actor relations

Seq	ActorToActorRelationship	Relation description
1.0	WouldLikeToKnow	Actor1 would desire to know Actor2 more closely
1.1	KnowsOf	Actor2 has come to be known to Actor1 through his, her, or its actions or position
1.2	KnowsByReputation	Actor2 is known by Actor1 primarily for a particular action, position or field of endeavor
1.3	KnowsInPassing	Actor1 has slight or superficial knowledge of Actor2
1.4	HasMet	Actor2 has met Actor1 whether in passing or longer
1.5	AcquaintanceOf	Actor2 has more than slight or superficial knowledge of Actor1 but short of friendship
1.6	TalksWith	Actor1 and Actor2 talk together
1.7	FriendOf	Actor2 shares mutual friendship with Actor1
1.8	CloseFriendOf	Actor2 shares a close mutual friendship with Actor1
1.9	LostContactWith	Actor2 was once known by Actor1 but has subsequently become uncontactable
2.0	LivesWith	Actor2 shares a residence with Actor1
2.1	NeighborOf	Actor2 lives in the same locality as Actor1
3.0	AmbivalentOf	Actor1 has mixed feelings or emotions toward Actor2

Table 5.5 Sample actor-to-thing relations

Seq	ActorToThingRelationship	Relation description
1.0	Consumes	Actor consumes the thing or portions of the thing
2.0	CreatesOrProduces	Actor creates or produces the thing
3.0	Uses	Actor makes use of or derives benefit from the thing
4.0	HasInterestIn	Actor is concerned about the thing
5.0	OccupiesOrPossesses	Actor physically occupies or possesses the thing
6.0	Owns	Actor owns or has some property rights to the thing
7.0	Controls	Actor controls access or use of the thing

member, etc.) between actors and the relationships (owner, controller, occupier, etc.) between an actor and an environmental element.

Table 5.4 lists some of the actor-to-actor relations that are possible. Many of these relations are taken from a web site on vocabularies (Davis, 2010).

These actor-to-actor relations can be used in describing a network of actors. In a theater-level model, most of these relations will be between group-type actors, indicating superior-subordinate relations in hierarchies and friend-foe-neutral relationships between groups. However, in special cases, relationships between individual actors may be appropriate.

Table 5.5 provides an initial list of actor-to-environmental-element relations. In a theater-level model, the most important actor-to-thing relations will probably be those identifying which group-type actors own or control or occupy which geographical areas or parts of the infrastructure.

A Total Ontology

Both the situation-independent and the situation-dependent ontologies are required for a total ontology of unconventional conflict. Despite the efforts to ensure completeness (for modeling at the "theater" level) with consistency (appropriate granularity), it should be expected that further use of the ontology will reveal some additional needs.

New Context Diagram

Figure 5.20 on page XX provided the context diagram for the discussion of the situation-independent ontology. Figure 5.42 provides a revised context diagram that includes the internal semantic relationships that are part of the situation-independent ontology and adds the new structures that are required to capture the situation-dependent information. This new context diagram omits several connections, such as the ownership link between Actor and GTO Set, to allow for a compact figure; however, these links are implied.

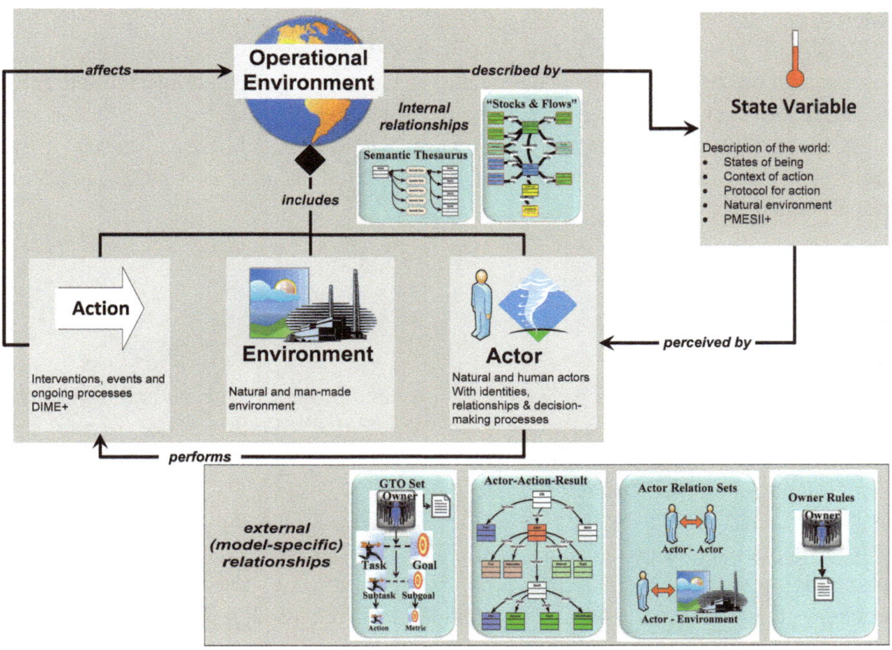

Fig. 5.42 Context diagram including situation-dependent parts

The complete ontology is a holistic description of the domain of unconventional conflict. That is, the ontology is "characterized by comprehension of the parts of something as intimately interconnected and explicable only by reference to the whole."

Adding to the Ontology

In the beginning, as new elements or metrics were identified, they were simply added to the appropriate parts of the ontology. Now, the process of adding to the ontology is much more complex due to the inter-relations that have been described earlier.

For example, suppose a new actor class is proposed. The first step is to see if that class already exists. There are too many classes to rely on memory of what already exists. Further, there may be an existing class, that is similar and a decision needs to be made about extending it, versus adding a new class. A new actor may imply the need for a new action (or vice versa). A review of the existing stocks-and-flows sets may reveal that the new element should be added to one of these. However, it might be that a new stocks-and-flows set is required and, perhaps, additional new elements may be needed to fill the roles in the new set. All the new elements must be checked for cascading connections and all need to be examined for connections to the sources.

If a new element is added, a new metric is going to be required—and the connection has to be defined. The new metric has to be connected into the semantic thesaurus and reviewed for connections to the sources.

If these additions are made in connection with a particular situation, all of the new elements and metrics need to be considered as entries to the situation-dependent structures.

Using the Ontology

Constructing the IW Ontology was not a "science project," with its creation as the sole goal. It was constructed to be used for DIME/PMESII models. Several use cases were developed to illustrate possible uses; two have already been implemented.

TRAC used the list of metrics in the IW Metrics Ontology (Hartley & Lacy, 2011) to plan for improvements to the IW Tactical War Game (TWG). TRAC personnel reviewed each metric (and its definition) to determine whether the concept it represented was already included in the model. Those metrics that were not already included were binned into three categories: those that should be included in the next year's model improvements, those that should be included in future improvements, and those that were not germane to the model's intended uses.

The TRAC users passed these results to the IW TWG modelers. Other model users can employ the Ontology in the same manner for any existing model in the unconventional conflict domain.

Hartley Consulting has also used the ontology to improve its VV&A tool, which was used to perform V&V of the HSCB Testbed and DEXES. These are discussed in Chapter 6.

In Chapter 7, we will describe the value of the ontology in building a model of unconventional conflict. We do this by sketching the process of building a simulation using the ontology.

Chapter 6
Verification, Validation, and Accreditation

Is this model useful?

"All models are wrong; but some are useful." This adage by the statistician George Box contains the rationale for verification, validation, and accreditation (VV&A) (Box, 1979). Together, they are meant to answer the question of whether a model is useful (for a given purpose). In this chapter, we define and discuss the terms and the practice of VV&A as it has applied to combat models and to models of unconventional conflict. In the process, we develop a prescription for its application.

Validation: The process of determining the degree to which a model is an accurate representation of the real world from the perspective of the intended uses of the model (Department of Defense, 2007).

Verification: The process of determining that a model implementation accurately represents the developer's conceptual description and specifications (Department of Defense, 2007).

Accreditation: An official determination that a model is acceptable for a specific purpose.

Together these are known as VV&A and V&V when accreditation is not being discussed. VV&A is required for DoD models; however, as they said in my math textbooks, "the details are left to the student."

V&V of Combat Models

I have personally participated in a number of V&V activities over the years. The first several involved combat models (Hartley, 1975, 1991, 2001d; Hartley, Kruse, Denson, Austin, & Breuer, 1989; Hartley, Quillinan, & Kruse, 1990; Hartley et al., 1994). Youngblood et al. describe a number of other V&V activities in the same time period (Youngblood, Pace, Eirich, Gregg, & Coolahan, 2000).

While the V&V for combat models is demanding, V&V of unconventional conflict models is challenging. Prescriptions for V&V of less exotic models abound.

D.S. Hartley III, *Unconventional Conflict*, Understanding Complex Systems,
DOI 10.1007/978-3-319-51935-7_6

Knepell and Arangno wrote an excellent book on V&V of general simulation models (Knepell & Arangno, 1993) and I have participated in articles about V&V of military simulations and their data (Hartley, 1997; Hartley & Whitley, 1996).

Formal VV&A of military simulations was the exception, rather than the rule, in the early days of military modeling. The concepts of VV&A existed, but were not well defined until a 1990 MORS Simulation Validation (SIMVAL) workshop created definitions that could be universally accepted. These definitions were officially instituted by the 1994 version of DoD Directive 5000.59 (Department of Defense, 2007).

Even with the promulgation of the definitions and the requirement for VV&A of military models, the practice was limited. V&V was regarded as hard—read, expensive—and left as the last thing to be funded and done. This was particularly true for combat models. Models of purely technical nature ("physics-based models"), such as missile fly-outs, were the exceptions, as data from testing could be gathered and compared to model predictions.

VV&A of Model Data

Originally, the accreditation of model data was known as certification. As late as 1996, there will still questions on what was meant by verification, validation, and certification of data and what should be meant by these processes (Hartley & Whitley, 1996). In models and simulations of that time, the model and the data were generally easily separable, often held in different files. The bulk of the data for combat models was produced by military organizations tasked with that responsibility. Certain data were "blessed" by the Services (Army, Navy, and Air Force) so that anyone wanting to model Service units was supposed to use the blessed data to describe that Service's capabilities.

As models and simulations became more complex, the "model" under consideration became more tightly tied to the scenario data. In some sense, this culminated in the generic agent-based model, where the "model" is defined almost completely by the data. The separation of the data from the model had become an academic distinction, with less value in actual practice. Accordingly, much of the V&V efforts with regard to data are subsumed in the model V&V.

Models that rely on HSCB data (e.g., unconventional conflict models) require data that are produced by non-military sources. This means that rather than controlling or influencing the production of data (through the military chain of command), model users are forced to understand the data that exist and cope with its problems. Often this requires reformatting and revising (to fit the model's assumptions) the data that are directly produced by the source. It also may require creating data. These problems were not uncommon in the old combat models, but comprise a much larger portion of the data activities for unconventional conflict models. Thus, the V&V of data, separate from the V&V of the model + the data, focuses on the nature of the original data and on the actions needed to make the data fit the model.

Developing V&V of Unconventional Conflict Models

V&V of unconventional conflict simulations is relatively new—it required someone to be interested in developing such a simulation. The SEAS (Chartuvedi, 2003) and DEXES models were probably the earliest, followed by DIAMOND and ISSM, all described in Chapter 3.

FAST

When DMSO began the FAST DIME/PMESII project, V&V was an integral part of the project. Each year there was a V&V plan, a V&V activity, and a V&V report.

The content of the verification and validation evolved over time. The first year of the project was consumed with identifying the parts that would make up the toolbox and testing the initial models to see if they would run separately and demonstrate some of their advertised functionality (Hartley, 2003).

We had to create a rational set of tests and devise a way of understanding the results of the tests. All of the test data were contained in spreadsheets, custom-built for the system and the tests. Once the tests had been agreed on in a "test definition" sheet, they were imported into a set of "test description" sheets. Figure 6.1 shows

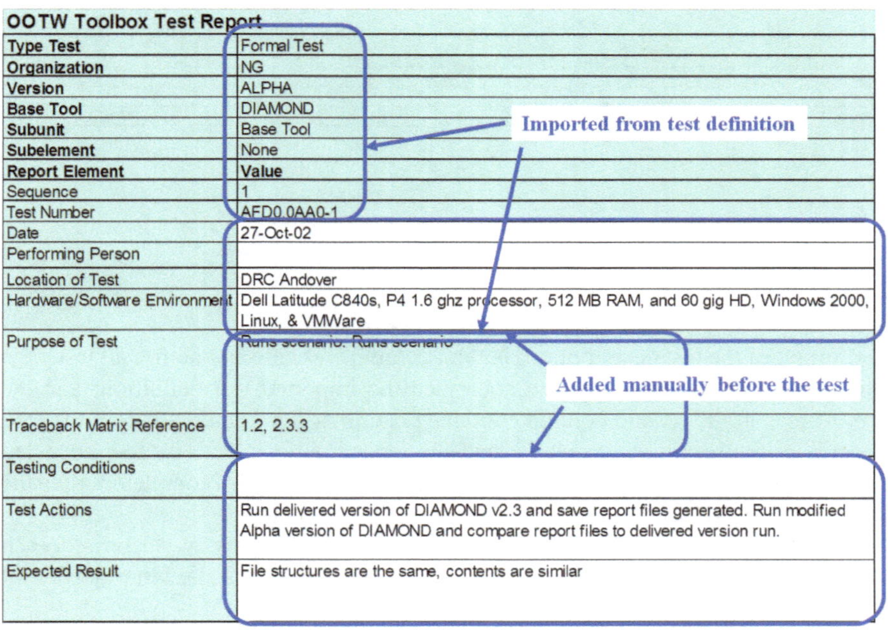

OOTW Toolbox Test Report

Type Test	Formal Test
Organization	NG
Version	ALPHA
Base Tool	DIAMOND
Subunit	Base Tool
Subelement	None
Report Element	**Value**
Sequence	1
Test Number	AFD0.0AA0-1
Date	27-Oct-02
Performing Person	
Location of Test	DRC Andover
Hardware/Software Environment	Dell Latitude C840s, P4 1.6 ghz processor, 512 MB RAM, and 60 gig HD, Windows 2000, Linux, & VMWare
Purpose of Test	Runs scenario. Runs scenario
Traceback Matrix Reference	1.2, 2.3.3
Testing Conditions	
Test Actions	Run delivered version of DIAMOND v2.3 and save report files generated. Run modified Alpha version of DIAMOND and compare report files to delivered version run.
Expected Result	File structures are the same, contents are similar

Fig. 6.1 FAST 02 test description format

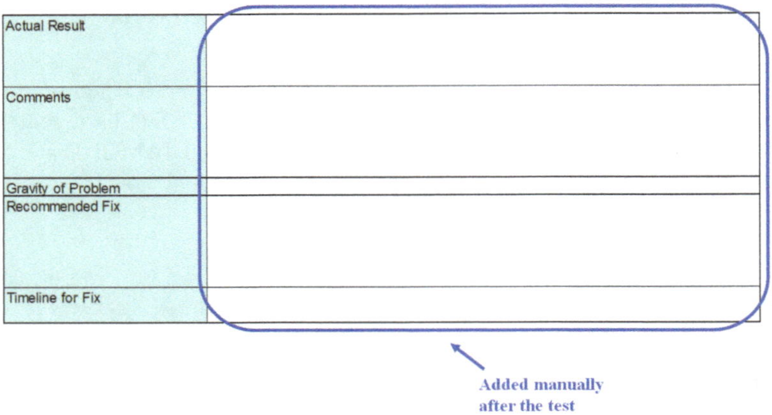

Fig. 6.2 FAST 02 test result format

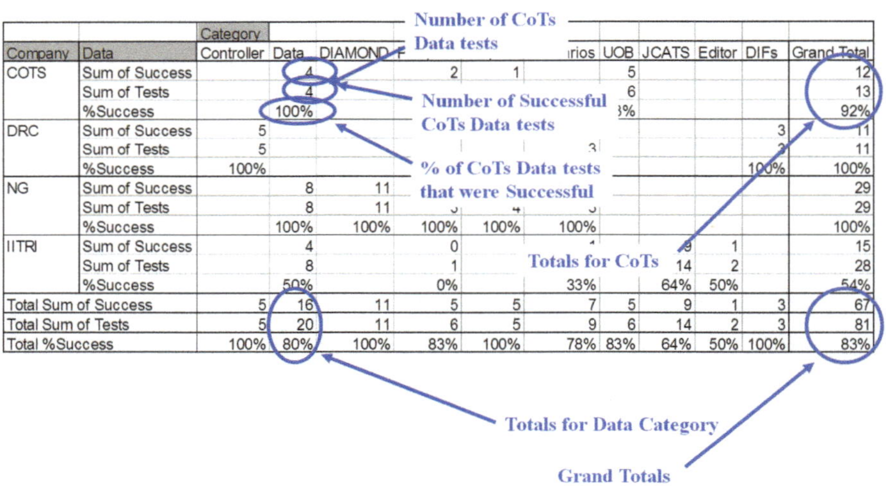

Fig. 6.3 FAST 02 sample summary results by performer

the format of the test description. The first column was constant across all tests; the blue areas in the second column contained the imported test definition; and the white areas in the second column contained data that were entered before the test, describing the testing environment and the test conditions, actions, and expected results. Figure 6.2 shows the format for the test results, to be completed after the test.

Figure 6.3 illustrates a sample summary results of the tests by performer (each contractor, such as CoTs, was responsible for one model or some other portion of the toolbox).

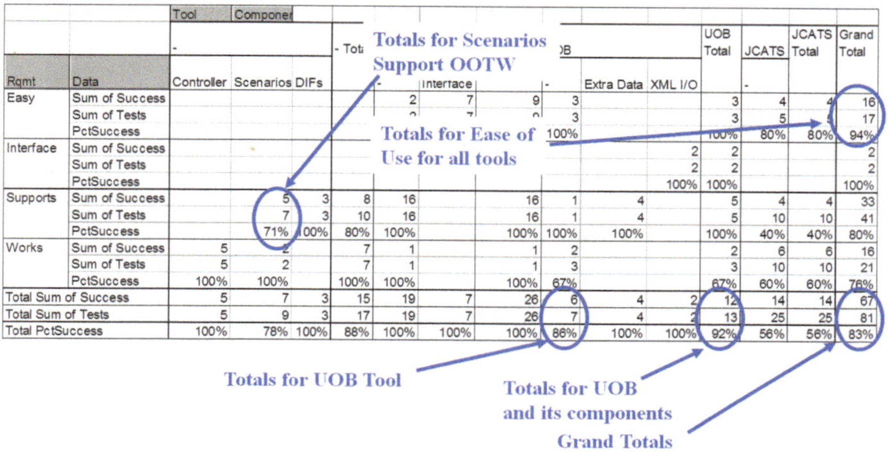

Rqmt	Data	Tool: Controller	Componen: Scenarios	DIFs	- Tot	-	Interface		- (…B)	Extra Data	XML I/O	UOB Total	JCATS -	JCATS Total	Grand Total
Easy	Sum of Success					2	7	9	3			3	4	4	16
	Sum of Tests					[illegible]	[illegible]	[illegible]	3			3	5	[illegible]	17
	PctSuccess								100%			[illegible]	80%	80%	94%
Interface	Sum of Success										2	2			2
	Sum of Tests										2	2			2
	PctSuccess										100%	100%			100%
Supports	Sum of Success		5	3	8	16		16	1	4		5	4	4	33
	Sum of Tests		7	3	10	16		16	1	4		5	10	10	41
	PctSuccess		71%	100%	80%	100%		100%	100%	100%		100%	40%	40%	80%
Works	Sum of Success	5	2		7	1		1	2			2	6	6	16
	Sum of Tests	5	2		7	1		1	3			3	10	10	21
	PctSuccess	100%	100%		100%	100%		100%	67%			67%	60%	60%	76%
Total Sum of Success		5	7	3	15	19	7	26	6	4	2	12	14	14	67
Total Sum of Tests		5	9	3	17	19	7	26	7	4	2	13	25	25	81
Total PctSuccess		100%	78%	100%	88%	100%	100%	100%	86%	100%	100%	92%	56%	56%	83%

Fig. 6.4 FAST 02 sample summary results by test requirement

Figure 6.4 presents the same data with a focus on how the models scored on satisfying the test requirements of ease of use (verification), proper interface functionality (verification), supporting OOTW (validation), and working as advertised (verification).

The 2003 version of the FAST Toolbox incorporated additional models, more ambitious interface requirements, and a more ambitious testing scenario (Hartley, 2004). Based on the previous year's experience, an explicit testing strategy was expressed.

The software engineering concepts, Alpha and Beta Test, were utilized. By "Alpha Test" we meant an early test and by "Beta Test" we meant a test of what we hoped was the final prototype (as opposed to pre-release software). It was important to make such distinctions since the Flexible Asymmetric Simulation Technologies (FAST) initiative in FY03 continued to build upon prototype software, which would not be ready for general release (in the commercial sense) even after a successful Beta Test.

Given the aforementioned clarification of terms, the Alpha Test addressed the question, "does the toolbox and all of its components do what we think they do?" The test items were verification tests, including regression testing to ensure that the process of fixing errors or enhancing utility did not introduce errors in areas that had previously worked. The emphasis was on finding failure points to be corrected prior to the Beta Test.

In the Beta Test, we revisited any failed tests from the Alpha Test; however, the focus shifted toward addressing the question, "do the toolbox and all of its components do what they should do?" Thus the test items included both verification tests and validation tests. The Beta Test Plan was coordinated with the Verification & Validation (V&V) Plan. The emphasis of the verification part of the Beta Test was on ensuring nothing would go wrong during user tests (following the Beta Test).

The administrative process of testing was similar to that of the previous year: custom spreadsheets were built to capture the tests, the results, and to analyze them. The test requirements were different, however: improve functionality (verification), integration with XML (verification), use of C4I data (verification), and OOTW use case (validation). Table 6.1 shows the comparison of the tests that were successfully performed by model and test requirement versus the number of scheduled tests. Table 6.2 shows the number of tests that were successful versus the number that were completed, by model and test requirement.

Naturally, there was another table (not included here) comparing the results of successfully completed tests to the planned tests. The point is that it was important to separate tests that were completed, but not successfully, from tests that could not be completed in order to determine where the problems were. For example, a test might not be performed because of the failure of a previous test, which was required to produce the inputs needed for the given test.

The 2004 version of FAST concentrated on using the toolbox to investigate alternate COAs for a particular scenario and to determine the capability of the toolbox to produce useful measures of merit for determining the best COA.

In the 2004 version of FAST, V&V responsibility and test responsibility were separated for the FAST Beta release. However, there were some clear overlaps that required coordination. For the most part, the V&V team (a group of military personnel not involved in the FAST project) performed validation while the test team performed verification and testing (which could inform both verification and validation) (Hartley, 2005b).

The testing was performed in the same manner as in previous years; however, additional questionnaires were designed for the V&V team to ascertain validation results. The emphasis was to evaluate the degree to which the FAST toolbox provided an accurate representation for a selected real-world scenario, from the perspective of its intended users, for planning, course of action (COA) analysis, and decision support.

The main conclusion was that DIAMOND and ISSM (Hartley, 2006a) did not exhibit any validity problems that would preclude their use in analysis or training. Their use for acquisition was not investigated. This should be regarded as a weak conclusion, as the examination should be characterized as "initial," not "thorough," and certainly not "definitive." This report was not an accreditation report. Rather it was a report on the status of verification and validation.

A secondary conclusion was that the two models showed some complementarity in their strengths and weaknesses. As a result, together they provided a more comprehensive representation of OOTW factors than either did alone. A degree of synergism was obtained.

Most of the available funds in FY 05 were earmarked for performing a classified study. Most of the rest of the funds were for testing/verification efforts (Senko, 2005). Therefore, the validation efforts were performed as adjuncts to these parts of the project (Hartley, 2005a).

The core of the FY 05 strategy was to take advantage of ongoing activities. The activities fell into two categories, actual use of the tools (FAST components) and

Table 6.1 FAST 03 beta test results: completion of tests

Company	(All)									
		Component								
Objective	Data	Controller	DIAMOND	ISSM	JCATS	Mine DB	Toolbox	UOB	XMT	Grand total
1. Improve function	Sum of tests	17.00	19.00	7.00	16.00	7.00	0.00	27.00	4.00	97.00
	Sum of desire	17.00	22.00	9.00	17.00	7.00	7.00	27.00	8.00	114.00
	Percent	100%	86%	78%	94%	100%	0%	100%	50%	85%
2. Integration w/XML	Sum of tests	3.00	3.00		2.00		0.00	4.00	1.00	13.00
	Sum of desire	3.00	3.00		2.00		1.00	4.00	2.00	15.00
	Percent	100%	100%		100%		0%	100%	50%	87%
3. Use C41 data	Sum of tests		2.00		6.00		0.00			8.00
	Sum of desire		7.00		6.00		1.00			14.00
	Percent		29%		100%		0%			57%
4. OOTW use case	Sum of tests		30.00	10.00	26.80		0.00	1.00		67.80
	Sum of desire		36.00	12.00	29.00		2.00	1.00		80.00
	Percent		83%	83%	92%		0%	100%		85%
Total sum of tests		20.00	54.00	17.00	50.80	7.00	0.00	32.00	5.00	185.80
Total sum of desire		20.00	68.00	21.00	54.00	7.00	11.00	32.00	10.00	223.00
Total percent		100%	79%	81%	94%	100%	0%	100%	50%	83%

Table 6.2 FAST 03 beta test results: success of completed tests

Company	(All)									
		Component								
Objective	Data	Controller	DIAMOND	ISSM	JCATS	Mine DB	Toolbox	UOB	XMT	Grand total
1. Improve function	Sum of success	16.75	18.60	7.00	16.00	6.90	0.00	27.00	3.80	96.05
	Sum of tests	17.00	19.00	7.00	16.00	7.00	0.00	27.00	4.00	97.00
	%Success	99%	98%	100%	100%	99%		100%	95%	99%
2. Integration w/XML	Sum of success	3.00	2.75		2.00		0.00	4.00	1.00	12.75
	Sum of tests	3.00	3.00		2.00		0.00	4.00	1.00	13.00
	%Success	100%	92%		100%			100%	100%	98%
3. Use C41 data	Sum of success		0.00		6.00		0.00			6.00
	Sum of tests		2.00		6.00		0.00			8.00
	%Success		0%		100%					75%
4. OOTW use case	Sum of success		29.80	10.00	27.00		0.00	0.98		67.78
	Sum of tests		30.00	10.00	26.80		0.00	1.00		67.80
	%Success		99%	100%	101%			98%		100%
Total sum of success		19.75	51.15	17.00	51.00	6.90	0.00	31.98	4.80	182.58
Total sum of tests		20.00	54.00	17.00	50.80	7.00	0.00	32.00	5.00	185.80
Total %Success		99%	95%	100%	100%	99%		100%	96%	98%

extended testing of the tools. Each category had different strengths. The actual use category was strongest in its requirement for extended use over time of a consistent set of inputs and the requirement for in-depth use of the tools. The extended testing category was strongest in its disciplined coverage of a wide range of tool functionality. Together, these two categories provided the opportunity to expose any weaknesses of the tools.

For ISSM, there were two categories of use, monitoring an actual OOTW and support to a study in which a simulation provides the OOTW state data. Both were included in the FY 05 Validation Plan.

The testing involved many of the same features as in previous years; however, the addition of Pythagoras, an ABM, added some novel features, and the need for simultaneous and sequential running of models on different computers added some novel test scheduling features (see Fig. 3.30, for a view of the linkages of the models) .

One of the most important differences in the FY05 experience was the requirement that the system be accredited for use in the ESS study (described in Chapter 3, Models, FAST System). This accreditation was provided by the J-8 Study Director, based on the V&V results.

The FAST experience provided lessons on the practicalities of performing V&V and accreditation as well as developing a philosophy of VV&A for unconventional conflict.

DARPA VV&A

As mentioned in Chapter 3, in late 2006 Alexander Kott of DARPA asked Stuart Starr, Albert Sciarretta, and me to develop a methodology for performing verification and validation on the COMPOEX PMESII model DARPA was developing (Hartley & Starr, 2010; Starr et al., 2007). We put together a team, collected a body of literature on VV&A prescriptions, VV&A practice, PMESII considerations, and simulation technology, and used the literature and our collective experience to create this methodology.

The team began the research activity by performing an in-depth review of the VV&A literature. That review confirmed that the history of VV&A of PMESII models was limited and the level of V&V had generally been inadequate. Furthermore, that review revealed that a PMESII model requires integrating multiple domains that often are based on competing theories of varying maturity.

DARPA was interested in both standard (baseline) V&V approaches and V&V approaches that would support rapid changes to COMPOEX, dubbed compressed and hyper-compressed V&V. Thus there were three situations to be addressed: baseline, compressed, and hyper-compressed. I will only discuss the baseline approach here.

In developing the baseline VV&A process, the team built on the latest thinking in VV&A approaches. This included an entrenched process that would be employed

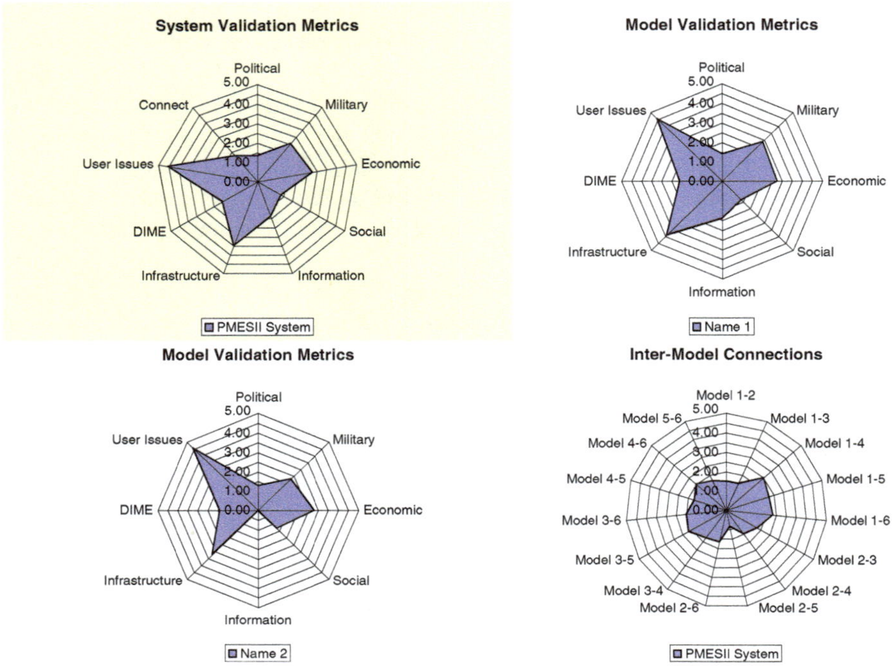

Fig. 6.5 V&V spider diagrams for PMESII models

throughout the life cycle of the model, the adaptation of a VV&A maturity model that was developed by DMSO, and the employment of a risk assessment and mitigation process to enhance communications with the user of the model.

As its first key concept for PMESII VV&A, the team developed systematic V&V metrics for PMESII models. The metrics were incorporated into a spreadsheet model that organized the tests and documented their results. These results captured the degree of success achieved in each test. However, with hundreds of tests covering the many areas of the system, simple combinations of the results tend to obscure rather than elucidate problems. To enhance the transparency of those tests, the team linked a set of "spider" diagrams to the spreadsheet (see Fig. 6.5). The "spider" diagrams supported visualization of multiple dimensions in a single chart and supported an overview and segmentation by each individual model. The metrics and their diagrams provided the V&V Team with deep insight into a model's strengths and weaknesses that emerged from the baseline VV&A process. The metrics and their diagrams also supported progress measurement as problems are fixed and are useful in conveying V&V status to the user.

In this figure, it is assumed that the system consists of multiple models, although only "Name 1" and "Name 2" are illustrated. The first spider diagram is for the system as a whole. The dimensions are each of the PMESII categories, plus DIME as a unit, User Issues (composed of some standard issues and any system-specific issues), and (inter-model) Connections. Each model has its own spider diagram, with the same dimensions, less Connections. "Inter-model connections" has its own spider diagram. The dimensions for this diagram are the connections between pairs of models (1-2, 1-3, 2-3, etc.).

Naturally, these charts could be decomposed further. The system and model charts could be decomposed into charts for each dimension, where each PMESII area has its own chart with the sub-areas as dimensions. The DIME area could be decomposed into the PMESII dimensions and the User Issues could be decomposed into the separate issues. The connection chart could be decomposed into the individually named connections for each pairing of models.

However, the purpose of the charts is not the proliferation of pictures, but a means of quickly identifying where there is weakness in the model. This identification helps in designing mitigations.

In each case the dimensions are scaled from zero to five, with zero indicating no information or complete failure (generally found where the model does not address the dimension at all). A score of five would indicate absolute validity, which would be rare in a PMESII model. The validation metrics are based on the conceptual models underlying the system.

As shown in Table 6.3, the definitions for the metric values depend on whether a component is being evaluated or whether an ensemble of components is being evaluated. The highlighted definitions are the ones judged most likely to be found in a PMESII model.

For coded models (conceptual models rendered into computer code), the V&V metrics are based on the following:

- A large (eventually exhaustive) set of PMESII variables and DIME functions that are relevant to the model's intended use;

Table 6.3 Validation metrics for conceptual models

Label	Component	Ensemble
5	Expresses fully validated theory, e.g., Newtonian physics with caveats on operations near light speed or in regimes subject to quantum effects	Expresses multiple Level 5 theories with fully engineered interfaces, e.g., fly-out model of ground-to-air rocket, involving, chemical reactions of propellants, ballistics, air flow, electronics, etc.
4	Expresses well researched theory involving considerable data checking and peer review, e.g., economic theory earning Nobel prize	Expresses multiple Level 4 or Level 5 theories with well researched interfaces, e.g., economic model ensemble used by Federal Reserve in setting U.S. interest and discount rates
3	Expresses theory supported by data and published in peer-reviewed literature	Expresses multiple Level 3, 4 or 5 theories with considerable peer-reviewed interfaces, e.g., some U.S. combat models
2	Expresses theory with rational basis, accepted by some experts as plausible (SWAG)	Expresses multiple Level 2 - 5 theories with plausible interfaces (SWAG)
1	Expresses a codified theory (WAG)	Expresses multiple Level 1 - 5 theories with codified interfaces (WAG)
0	Uncodified, mental model of uncertain consistency and completeness	Uncodified processes for connecting models of uncertain consistency and completeness

- The Conceptual Model (CM) validity metric (1–5), described above, for each variable and function;
- Suitability for the given use (a fractional value) for each variable and function, based on
  - Granularity of model matched to granularity needed for the use,
  - General match of model to use,
  - Correct direction of change, and
  - Credible order of magnitude change; and
- The PMESII components are also measured by the internal connection validities within each model of the system of models.

The connections among the models are measured for each relevant model-to-model pair by suitability for the given use (a fractional value) for each pairing, based on

- Granularity of model matched to use,
- General match of model to use,
- Correct direction of change, and
- Credible order of magnitude change.

The User Issues are measured for each model and for the system of models by

- Ease of use,
- Speed of use, and
- General fitness of the model or ensemble to the particular use.

The team determined that the second key concept for PMESII VV&A was what it called entrenched VV&A. V&V is conducted by asserting functionalities of the model and testing the assertions. In entrenched VV&A, there are four types of testing: developmental, triggered, periodic, and model use testing. Additionally, there is an accreditation process.

Figure 6.6 illustrates the elements of entrenched VV&A. The first arm of the Type of V&V decision box is development testing, which emphasizes that this testing must be considered to be part of V&V. This includes standard model-creation processes: design walkthroughs, code review, debugging, etc. It also includes standard model-acceptance processes: alpha testing, beta testing, acceptance testing, etc. If the model is a system of models, it also includes multi-model testing.

The second arm of the decision is periodic testing. The point is that a model requires a testing campaign of defined periodic testing because no model of any size can be completely tested during any test event. Periodic testing permits additional testing to improve the knowledge concerning the model (as long as good records are kept). These tests include different data sets and tests of different functionality along with regression testing to ensure that things that used to work still work.

The third arm of the decision is triggered testing. Triggers that should lead to new tests include the identification of a major error in the model, the addition of

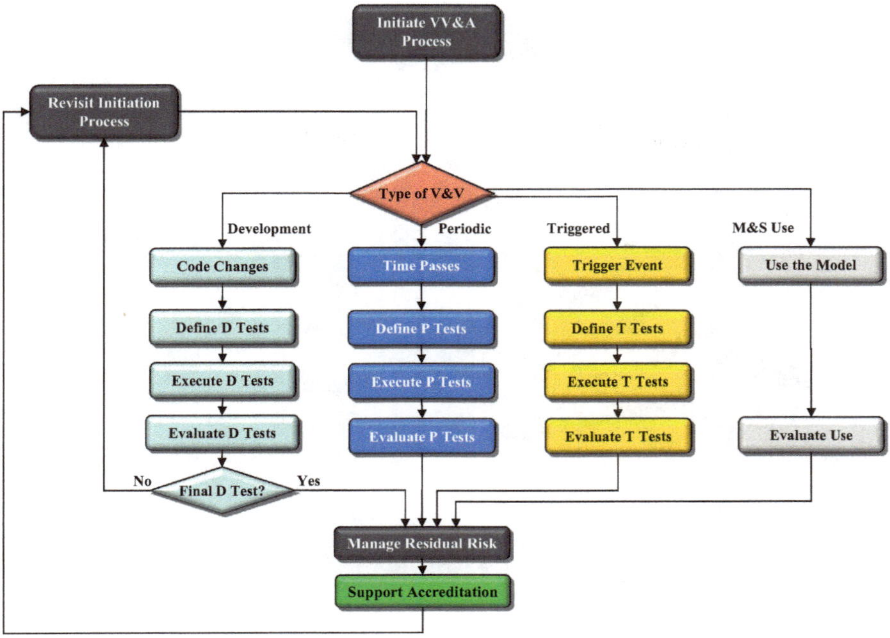

Fig. 6.6 Simplified VV&A process diagram

new processes to the tool, changes to existing processes, and changed tool environment processes.

The fourth arm of the decision is use of the model. This is not the execution of new tests; rather the execution of the model in its use is a test. The results of the use of the model must be recorded, both bad (perhaps leading to additional work on the model) and good, which adds to the confidence in the model.

As shown in the figure, following each arm of the decision is an activity to mitigate any risks that are discovered. This can include program changes or notes to avoid certain situations in model use, or work-arounds that have been devised.

Finally, there is the accreditation process, which is designed by the accrediting agent. To support this, the V&V process will generate verbal results of one or more of the following types:

- Absolute (complete pass, partial pass, failure: e.g., with respect to domain coverage),
- Relative (better than previous standard, equal, worse: e.g., with respect to resolution level) and
- Suitability for use (ease of use, applicability to problem).

It may also generate caveats, such as:

- Basal—sufficient for instantiation, not for actual use;

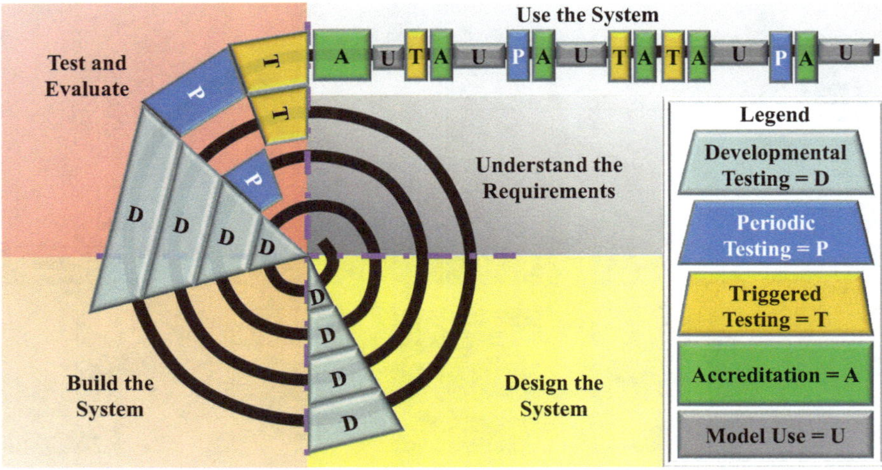

Fig. 6.7 Entrenched VV&A lifecycle

- Generic—sufficient for generic scenario use;
- Particular—sufficient only for particular scenario;
- Domain restriction—for example, if the model does not represent the effect of opium on the economy, it is not applicable for the Afghanistan AoR; and
- Problem set restriction—not good for predictive or comparative modeling.

Figure 6.7 illustrates the entrenched VV&A process as part of the entire model lifecycle. The left side of the figure stands for a spiral development process, cycling among "understanding the requirements," "designing the system," "building the system," and "test and evaluate the system." The light blue wedges with the "D" label represent the developmental testing. The blue wedges with the "P" label represent the periodic testing. The yellow wedges with the "T" label represent triggered testing. The green boxes with the "A" label represent accreditation. During the use of the system, there will be occasional triggered testing, scheduled periodic testing, and (usually aperiodic) accreditations. Each of these pairs of testing and accreditation has a mini-development cycle attached to represent the possibility of design changes resulting from the testing. This figure does not include the evaluation that occurs with each use, as it is assumed.

Entrenched VV&A was the second key concept of PMESII VV&A. The third key concept for PMESII VV&A was risk management and mitigation.

Model building and use entails risks. Figure 6.8 shows a generic diagram of model building and use and identifies the general risk points:

1. Using real-world data to create the conceptual model;
2. Using a proxy for the real world/perception of the real world to create the conceptual model;
3. Using the conceptual model to create the coded model;
4. Using real-world data to create the data for the model;

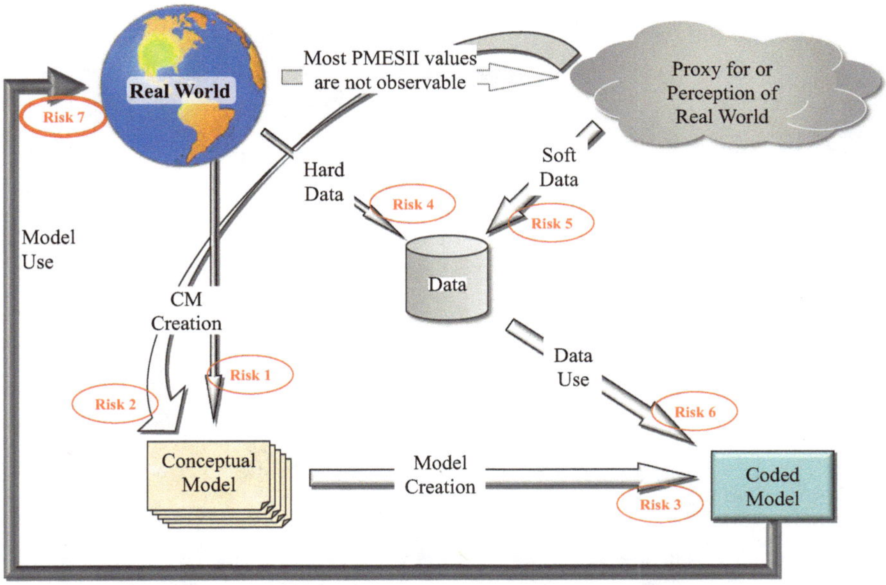

Fig. 6.8 Modeling risks

5. Using a proxy for the real world/perception of the real world to create data for the model;
6. Using the data to run the model; and
7. Using the model output to make inferences about the real world.

The V&V process should mitigate these risks (Fig. 6.9).

- The conceptual model (CM) validation plus theory, data and SME validation mitigate risks 1 and 2;
- Model verification mitigates risk 3;
- Data V&V mitigates risks 4, 5, and 6;
- Model validation plus theory, data, and SME validation mitigate risk 7.

To aid in risk assessment, the team modified the Youngblood Validation Process Maturity Model (VPMM) (Youngblood, 2005). Figure 6.10 shows the PMESII VV&A Maturity Model. The text in red is what has been added to account for the nature of PMESII models. The portion outlined in blue (levels 0 and 1) represents the probable maturity level of models at the time of the project. The portion inside the green outline (levels 2 and 3) represents the probable best achievable level.

The team used the maturity model to develop a matrix of risks and mitigations for the following items at each maturity level.

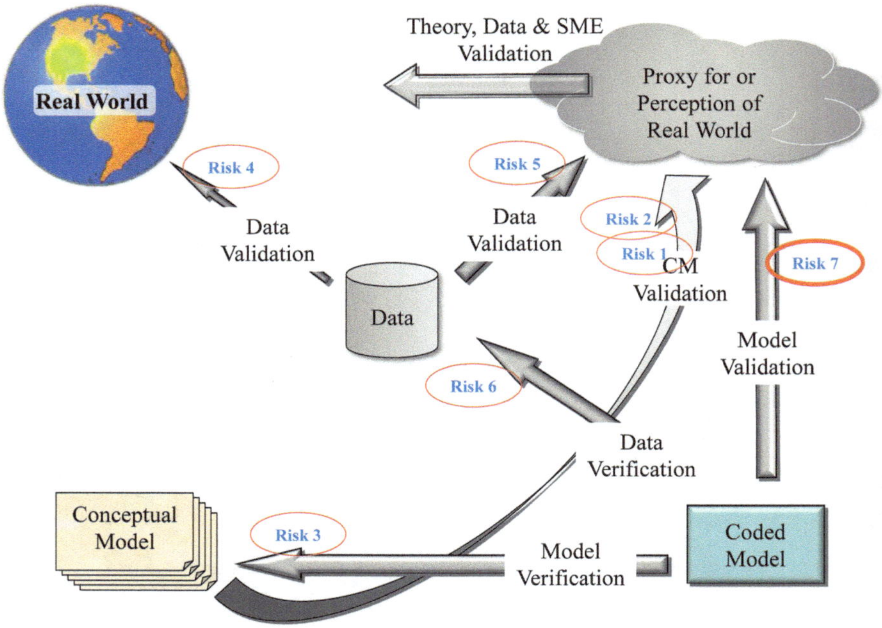

Fig. 6.9 Mitigating risks through V&V

Level	Validation Criteria	Referent	Conceptual Model	Development Products	Simulation Results	Documentation Consistency	Testing Standards	Configuration Management Standards
0	none derived	none chosen	none constructed	verified enough to support development	not validated at all	little or no documentation	no standard practices	no standard practices
1	represented by SME opinion	represented by SME opinion	none constructed	verified enough to support development	validated by SME observing simulation results	little or no documentation	no standard practices	no standard practices
2	defined from user statements as entities represented, their attributes & the dependencies among them	represented solely by SME opinion, based on intended use of the model/system	validated against the validation criteria by the SME	verified against the conceptual model inventory	validated by SME against the validation criteria	standard documentation produced, but limited in scope	standard practices with developer performing the V&V tests	manual configuration management, but not standardized
3	defined from user statements as Level 2 terms plus attribute ranges, domains & errors	derived from a single source, based on intended use of the model/system	validated by objective party from validation criteria & referent	verified against the conceptual model	validated by objective party from validation criteria & referent	standard documentation produced, with nearly complete scope	standard practices with someone other than the developer performing the V&V tests at acceptance or accreditation time	standardized manual configuration management
4	defined from user statements as Level 3 terms plus confidences	sampled from multiple independent sources, based on intended use of the model/system, & correlated statistically with estimates of uncertainties	validated by objective party from validation criteria & referent; analyzed to suggest results sampling space & estimate the confidence associated with the sampling	verified against the conceptual model; provides information to guide results sampling & estimate the confidence associated with that sampling	sampled from guidance developed from CM & verification analyses; validated by objective party from validation criteria & referent	standard documentation produced, with complete scope and database	standard practices with a disinterested party performing the V&V tests at acceptance or accreditation time	standard computer based configuration management
5	Level 4 terms formally derived from user statements using causality arguments	rigorously derived from multiple independent sources, based on intended use of the model/system, & characterized statistically with estimates of uncertainties	formally stated & validated automatically from validation criteria & referent: analyzed to define results validation sample space	verified against conceptualo model & used to define results validation sample space & the confidence associated with that sampling	automatically sampled from guidance developed from CM & verification analyses; validated automatically from validation criteria & referent	standard documentation produced, with complete scope and database and complete retrieval system	standard practices with a disinterested party performing the V&V tests at acceptance or accreditation time	standard computer based configuration management with check-out, check-in safeguards

Fig. 6.10 PMESII VV&A maturity model

- Methodology
  - Intended use
  - Concept of Operations (CONOPS) maturity
- Intellectual Capital
  - User experience
  - SME availability
  - SME quality
- Data
  - Particular Data

    Bad input data
    Misinterpreting results
  - General Data

    Wrong format
    Content nature
- Tool
  - Individual Model

    Model is biased
    Model is incomplete
    Model is wrong
  - Model Interactions

    Interactions are not traceable
    Interaction is biased
    Interactions are incomplete
    Interaction is wrong
  - General Tool

    Maturity of component
    Maturity of composed M&S
    Prior V&V
    Configuration management

Despite all mitigation processes, in general, there will remain residual risks. Part of the VV&A process is the communication of risks, mitigations and estimated residual risks to allow users to balance the risks against the benefits of using the model.

The DARPA V&V project developed three important concepts: V&V metrics for PMESII models, entrenched VV&A, and PMESII risk management and mitigation.

DIME/PMESII VV&A Tool

In 2008, following the conclusion of the DARPA VV&A project, Hartley Consulting built version 1.00 of the DIME/PMESII VV&A Tool (currently at version 1.52). This tool was based on the experiences of the FAST Toolbox VV&A and the results of the DARPA VV&A report. Its features include the following:

- Persistent database to retain the VV&A records of a system, supporting entrenched VV&A;
- VV&A based on system and model versions, not just names, supporting the concept that changes in a model imply changes in its VV&A status;
- Provision of test event creation, individual test creation, test event scheduling, and recording of test results, with configuration management of the tests, building on the experience of the FAST Toolbox VV&A, but eliminating the need for custom spreadsheets;
- Support for the full array of entrenched VV&A testing: developmental, periodic, triggered, and system use;
- A full PMESII-based metric for both the conceptual models and the coded models of a system, with explicit model theoretical bases, as described in the DARPA report;
- Risk analysis, mitigation, and residual risk identification. included as integral parts of the test results reporting process;
- Extensive output reports including overview charts and text reports suitable for reporting to sponsors and detailed charts and text reports suitable for defining actions needed at the software programming level; and
- Provision of accreditation reporting, supporting various types of accreditation, and caveats.

PMESII Metric: The PMESII metric feature deserves amplification. The first item of concern when performing V&V of a model of unconventional conflict is its coverage of the DIME/PMESII domain. Before investigating how well the model represents unconventional conflict, it is important to identify what it represents. The purpose of the model drives the needed coverage. For example, an economic model must cover at least some parts of the Economic metrics of DIME/PMESII and an infrastructure model must cover at least some parts of the Infrastructure metrics of DIME/PMESII. A model that purports to be a full model of unconventional conflict should address all DIME/PMESII metrics.

The DIME/PMESII VV&A Tool provided one source for the creation of the Ontology (described in Chapter 5) and has been revised based on the completion of the ontology (Hartley, 2015b). [More details can be found on the Hartley Consulting website (Hartley, 2008).] In order to determine the PMESII coverage of a model, it is critical that the test for coverage is based on a correct listing of what should be covered. The revision of this tool based on the ontology included a revision of the PMESII variables used in performing V&V. Figure 6.11 illustrates this revision and the resulting correspondence between the 408 tool PMESII metrics

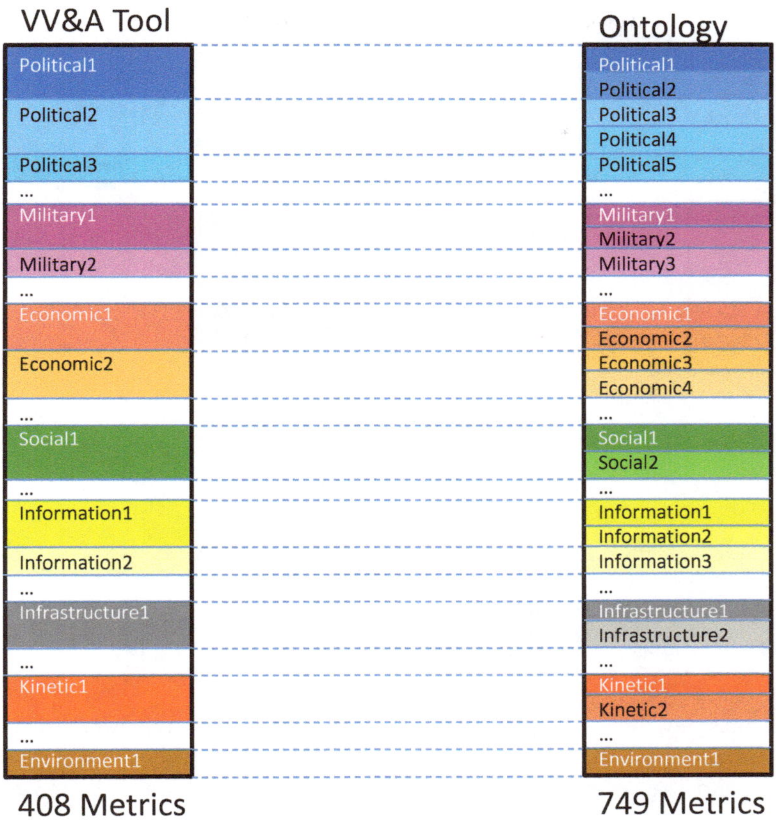

Fig. 6.11 VV&A tool and ontology metrics correspondence

and the 749 metric classes of the ontology. In performing V&V of a model a compromise is necessary between the precision gained by including each of the ontology metrics and the time required to check the coverage of each metric by a given model. The judgment was that a number of the ontology metrics could be reasonably combined because of their similarities. The result was a reduction of almost 50%, leaving an uncomfortably large, but manageable, number of checks, while retaining most of the precision of the ontology.

The second item of concern is the validity of the conceptual model(s) upon which the computer model is based. Each covered DIME/PMESII metric will depend on something, ranging from a fully validated theory to code that the programmer created to connect the element to the rest of the model (Table 6.3). In Fig. 6.12, the model has been examined to see which PMESII metric is covered (yellow highlights) and each of these is connected to the theory (or theories) that this model uses to assign values to the metric. (The model's metrics will have their

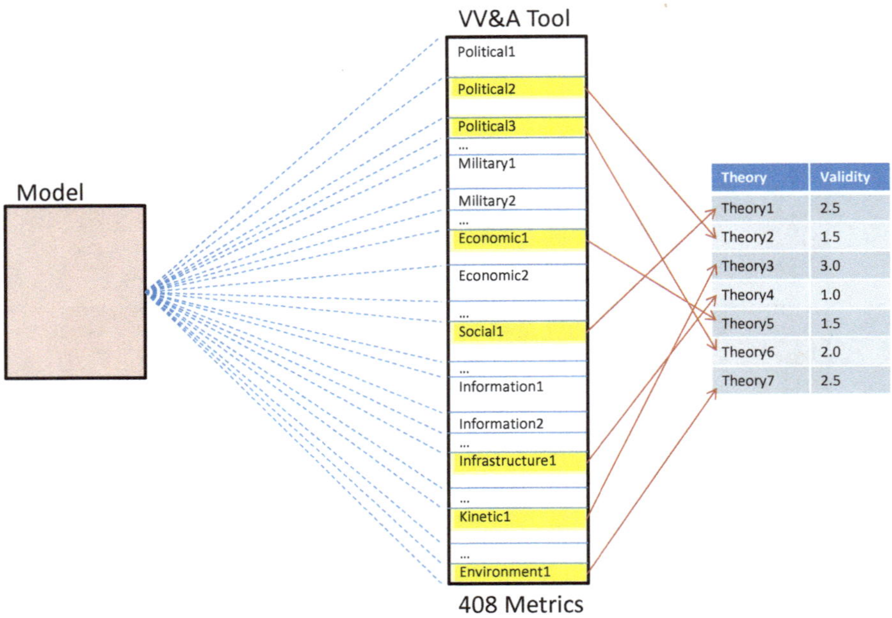

Fig. 6.12 Connecting covered metrics to theories

own names, not those of the tool PMESII metrics; however, this is not material to the V&V process.)

The third item of concern is the quality of implementation of the conceptual model (verification). Together with the validity of the conceptual model, this yields an estimate of the validity of the implementation of each DIME/PMESII metric.

Spider charts, such as shown in Fig. 6.5, are used to show the coverage and estimated validity of a model. For example, the Name 2 sub-model in the figure shows no coverage of the Information PMESII metrics. This may indicate a problem or it may indicate that Information is covered in a different sub-model and is not relevant for Name 2.

Process Flow: Figure 6.13 illustrates the process flow in using the Tool. The blue boxes indicate general preparatory processes. The orange boxes indicate preparatory processes for the dynamic testing, the testing requiring running the system. The yellow boxes indicate actual testing, both dynamic testing and static testing (the testing that involves examination of the conceptual model, the coded model, and the data). The green box is the test evaluation process and the pink box is the accreditation process.

System Definition: All of the processes are important; however, the system definition process is critical. Without knowing exactly what has been tested and accredited, all of the work of performing the tests is useless. Figure 6.14 illustrates the philosophy of system definition. The system is decomposed into models and the models are decomposed into modules. (A stand-alone model with no modules would be listed as the system, with a single model [itself] and a single module

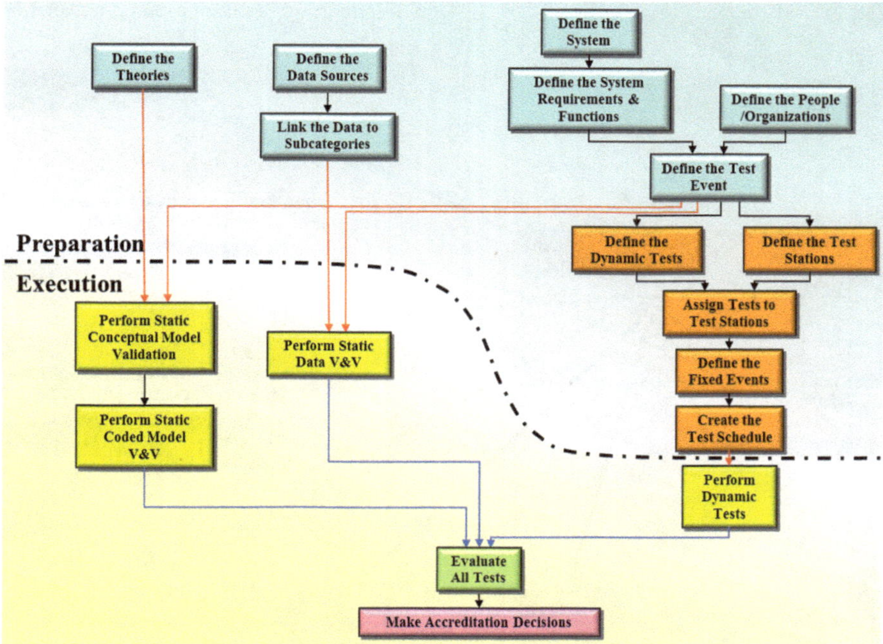

Fig. 6.13 DIME/PMESII VV&A tool process flow

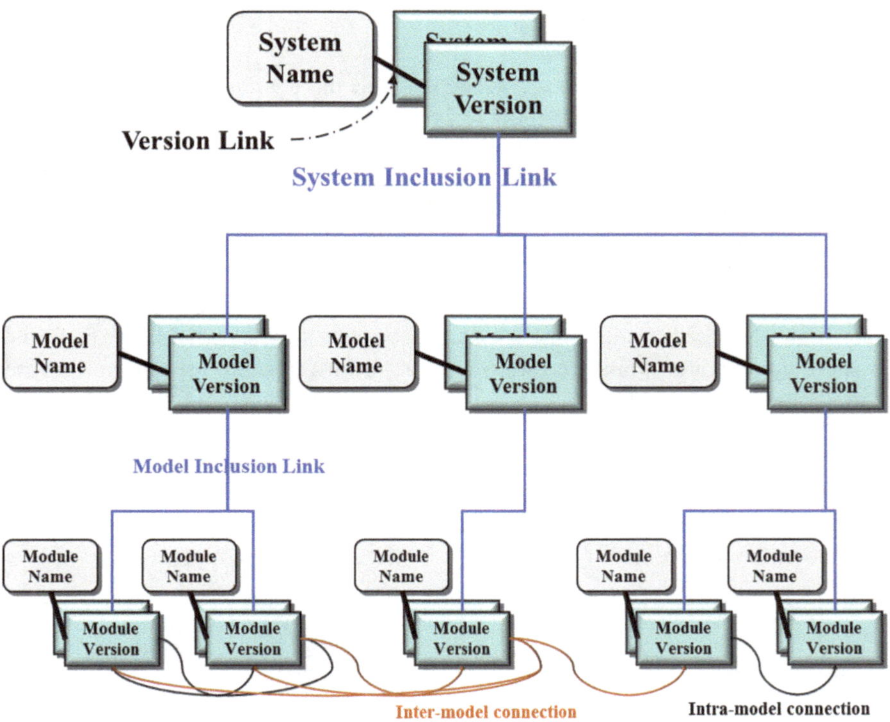

Fig. 6.14 System definition philosophy

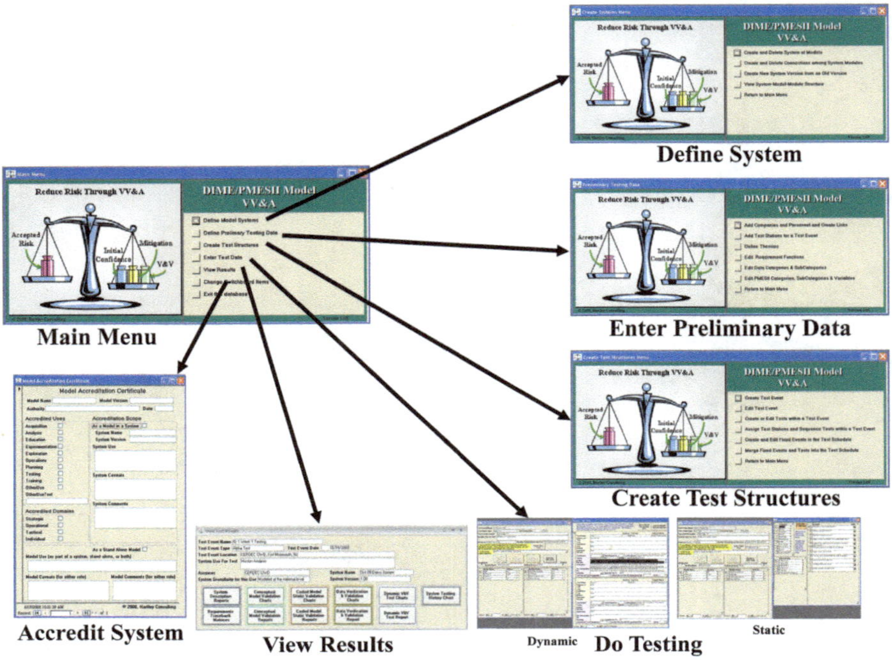

Fig. 6.15 DIME/PMESII VV&A tool menus

[itself].) The system, each model, and each module has a unique identifying version number. If the version of any module is changed, the model must have a new version number and the system must have a new version number. All connections between models are represented at the module level. The Tool maintains all records tied to the version numbers. If a new version of the system is being tested, but only one module has been changed, the data about the other modules can be retained.

Ease of Use: The DIME/PMESII VV&A Tool has been built for ease of use (to the extent possible). Figure 6.15 illustrates some of the menus of the tool. These menus are designed to allow the user to follow the flow shown in Fig. 6.13 and create the preparatory data, perform the tests, print the output reports, and record accreditation results.

The DARPA V&V project developed three important concepts: V&V metrics for PMESII models, entrenched VV&A, and PMESII risk management and mitigation. The DIME/PMESII VV&A Tool emphasized the need for system configuration management, the importance of identifying the theories used in the system, and added the requirement for an explicit VV&A management process.

HSCB Testbed

The HSCB Testbed was described in Chapter 3. In 2009, Hartley Consulting joined the project team specifically to incorporate V&V into the year's testing event

(Hartley, 2009). The early version of the DIME/PMESII VV&A Tool was used to support the V&V process (Hartley, 2010a).

DEXES Conceptual Model Validation

The DEXES II model was also described in Chapter 3. In 2011, Hartley Consulting was hired to support the U.S. Southern Command (USSOUTHCOM), its contractor, Northrop Grumman Corporation (NGC), and its sub-contractor, the University of Colorado—Denver, in their verification and validation of the DEXES-II computer system. In particular, Hartley Consulting was asked to support the validation process at the conceptual model level using a later version of the DIME/PMESII VV&A Tool (Hartley, 2012).

Other V&V Developments

The discussions above represent projects in which I was personally involved. This concentration is not meant to imply that these were the only projects of significance in the VV&A field nor that other projects did not influence the ones described above. Concurrent with these projects, others were doing significant work in the field.

One of the most influential workers in the field has been Paul Davis of RAND. In 2013, he wrote about basic issues in validation of HSCB and DIME/PMESII models (Davis, 2013). He said the paper is,

> . . . about validation and usefulness—complicated issues when dealing with social-science models. I discuss these complications and offer provocative suggestions, which include recognizing: the range of purposes for which the models are used, that validation should reflect the style of inquiry, that both theoretical and empirical information should be used for validation, that empirical testing should be theory-informed and theory-building oriented, and that more diverse types of empirical information should be used. Finally, I see value in validating the qualitative structure of uncertainty-sensitive systemic models. These can then be used for exploratory analysis in search of conclusions robust across uncertainties.

Review of VV&A of Unconventional Conflict Models

The process of developing a practical VV&A process for unconventional conflict models has required years of experimentation with methodologies, research into the literature of VV&A, application of the process on different models, and revisions based on work with ontologies. The DARPA V&V project developed three important concepts: V&V metrics for PMESII models, entrenched VV&A, and PMESII risk management and mitigation. The DIME/PMESII VV&A Tool emphasized the need for system configuration management, the importance of identifying the theories used in the system, and added the requirement for an explicit VV&A management process.

This development experience has shown that there are differences between the process for combat models and that for unconventional conflict models. However, there are also many commonalities.

The first commonality is that the VV&A process must be entrenched for both types of models. V&V includes all testing from model conception through model retirement, where the VV&A records are maintained in case the model is later brought out of retirement. To be useful, entrenched VV&A requires maintaining records of all testing throughout the life of the model. This brings up the second commonality: strict version configuration control is required to support the VV&A process and must be attached to the testing results.

The major difference between the VV&A processes lies in the metrics required for assessing validity. Unconventional conflict models rely on the use of social theories in determining the results of actions, whereas conflict models (at least those in the past) rely more heavily on physical theories, with stronger validity. This requires both an emphasis on PMESII metrics and the understanding that unconventional conflict models will require much more care in their use and assessments of their results.

In particular, the assessment of the theories used in a model becomes central to VV&A for models of unconventional conflict. See the discussion in the DIME/PMESII VV&A Tool, above, for a more detailed description of coverage of the metrics. Although risk assessment and mitigation—and the understanding of residual risks—are common to both cases, the likelihood of larger risks is inherent in unconventional conflict models.

Chapter 7
Modeling Unconventional Conflict

Everything should be made as simple as possible, but not simpler.

In this chapter, we will describe the value of the ontology in building a model of unconventional conflict. We do this by sketching the process of building a simulation using the ontology. This sketch identifies how each of the parts of the ontology supports model building. However, it is not a complete blueprint and is not meant to be one.

Designing the Model

The first item on the agenda in building a model is to define its purpose: Will it be used to support training? Will it be used to predict/forecast/explore the future? Will it be used to compare alternative tactics and strategies or alternate force compositions? The answers to these and similar questions will help define the needed span and granularity of the model, as well as the type of model that is needed.

We will assume that the model will need to be a simulation that is roughly of theater span and with granularity that matches the ontology described in Chapter 5. Otherwise, a different ontology would be needed. As a first step in the description, we will assume that the desired simulation will cover all possible unconventional conflicts. In the "Culling the Simulation" section, we will discuss how the ontology is useful in reducing the coverage for modeling particular situations.

Figure 7.1 illustrates the steps in using the ontology to design a model. Each of these steps is described in a subsection below.

D.S. Hartley III, *Unconventional Conflict*, Understanding Complex Systems,
DOI 10.1007/978-3-319-51935-7_7

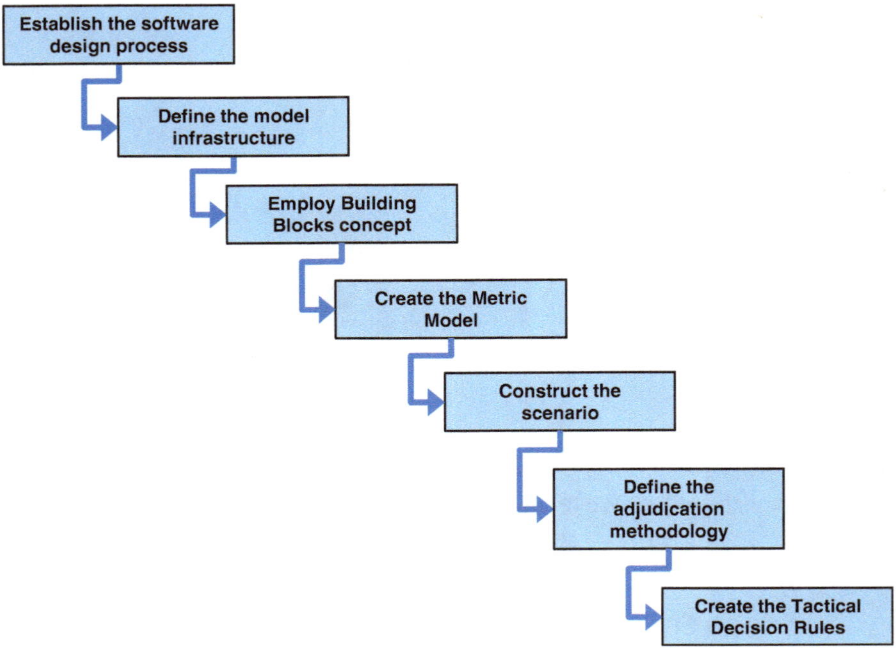

Fig. 7.1 Using the ontology to design a model

Software Design

With the existence of the holistic ontology of unconventional conflict, it is possible to define the majority of the requirements at the beginning of the project and to do the majority of the design after that. As a practical matter, it will still be wise to have cycles of requirements, design, coding, and testing (part of verification and validation); however, the requirements and design portions can be smaller parts of the cycle. This is illustrated in Fig. 7.2.

The ontology was described in Chapter 5. The design process using the ontology (with notes on the implementation) is described below.

Infrastructure

Coding the logic of a model is more fun than creating the infrastructure that supports the model; however, that infrastructure is required and must be created (Fig. 7.3).

Language: Because, the ontology is based on a class-object paradigm, an object-oriented language, such as C^{++} or Java, is the natural choice for building a model.

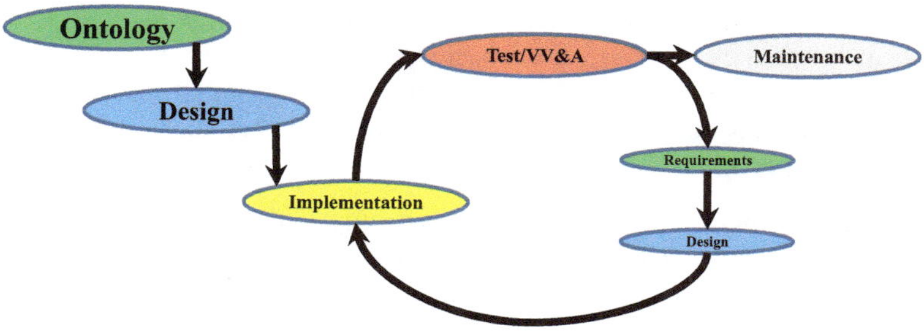

Fig. 7.2 Ontology-based software design

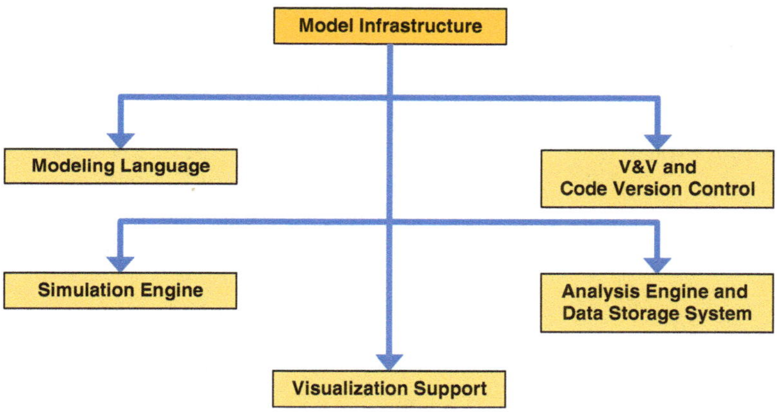

Fig. 7.3 Elements of the model infrastructure

However, although the multiple inheritance property may encourage the use of C^{++} , Java can implement most of multiple inheritance functionality through the use of interfaces.

Verification and Validation and Code Version Control System: V&V must be an integral part of the system and will require appropriate infrastructure, such as strong version control of the code. That infrastructure will be discussed in a later section.

Simulation Engine: The choice of a simulation model means that a simulation engine will be required. This engine can be a time stepped or discrete event engine. However, given the time span of many years for many unconventional conflicts in the past, the bias is toward discrete event simulation. The examination of short-timeframe vignettes can be accomplished with a time-stepped simulation; however, much longer time frames require the variable time resolution of discrete event simulation to hold model run-times down to reasonable levels.

Deterministic models have proved to be easier to explain to decision-makers than have stochastic models. And they have the advantage of requiring only a single run for a given scenario. However, unconventional conflict is inherently uncertain because it so heavily involves human choices and our poor knowledge of the theoretical underpinnings of modeling human behavior both strongly suggest that a stochastic model is the better choice. This implies the need for a well-thought out scenario input and output control structure.

Analysis Engine and Data Storage System: Whatever the choice between deterministic and stochastic modeling, a thorough output analysis structure will be required. This includes both the analytical logic and the storage controls and logic to support storing and retrieving the output data. In the stochastic modeling choice, a more complex structure will be needed to handle the analysis of multiple runs of each scenario and its variants.

Visualization: The model will need a strong geographical display capability. Once the model has been debugged for a particular scenario, the geographical display will need to be turned off to increase the run speed. However, the display will be critical in the debugging process to examine anomalies. The geographical display will also be required for explaining results. The model will also need a strong non-geographical display capability to help understand the differences in the results among the scenario variants and to help understand what factors are driving the results.

Building Blocks

The element and metric classes and their connections provide the building blocks for creating an object-oriented simulation based on the ontology (Fig. 7.4). Converting the large number of ontology classes to programmed classes will entail considerable effort; however, that work will not be technically difficult.

To the extent that the ontology is complete (for modeling), all of the needed building blocks for the model will be present in these classes. For example, the 102 actor classes, the 208 environment classes, and the 385 action classes serve as templates for all the instantiated active entities, passive entities, and actions, respectively, in a model. The 749 metric classes serve as templates for all the instantiated state variables.

The properties that have been defined (Fig. 5.30, page XX) populate the attributes of the element objects. The metrics and these attributes will provide a good start in defining the data input needs and the data outputs. There will be additional connections that will have to be made to the geographical display infrastructure and to the input and output structures. The element categories and the PMESII categories will provide some useful summarization features.

The next step in creating the building blocks is to implement the other situation-independent ontology structures into the code (Fig. 7.5). For example, the stocks-and-flows sets define some of the natural connections among the previously programmed classes and the semantic thesaurus will be useful in connecting the metrics into summary metrics, depending on the needs of the situation being modeled.

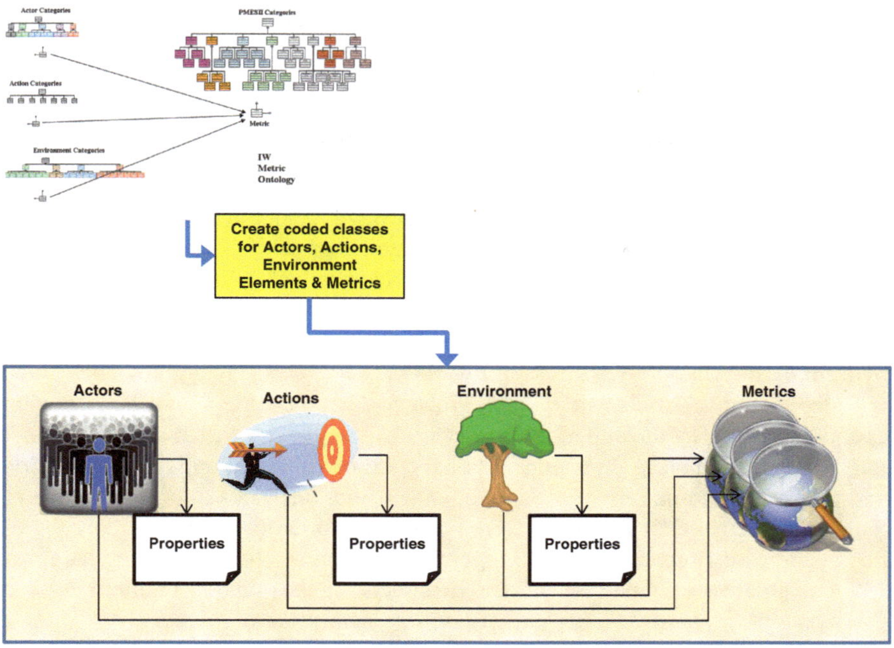

Fig. 7.4 Element classes as building blocks

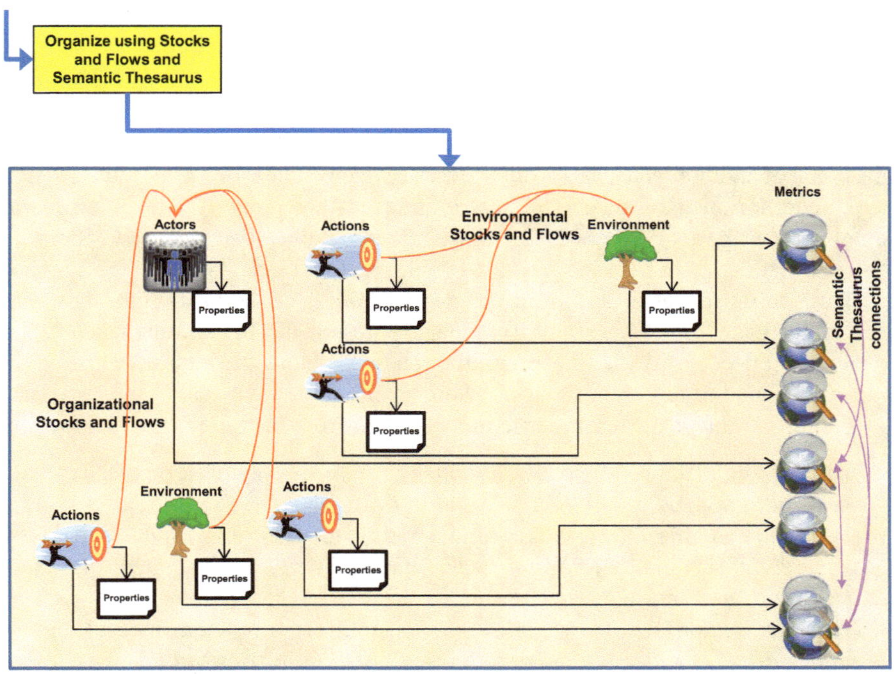

Fig. 7.5 Building blocks continued

Considering Perception Versus Ground Truth

Some models maintain only one set of state variables and some models maintain multiple sets. In a model with only one set of state variables, all actors know the actual state of the variables to which they have access. Thus actor A may not know everything that actor B knows; however, everything he knows accurately represents the state of the model—"ground truth." On the other hand, the model may maintain a set of state variables for each actor (or grouping of actors). The values of these state variables contain the perception of the actor, which will, in general, not be ground truth.

There are pros and cons for limiting the actors' knowledge of ground truth, either by limiting the variable set access or by allowing for false perceptions. The choice will depend on the use of the model. However, it should be noted that allowing complete access to the model's ground truth is equivalent to assuming perfect intelligence on all sides. In unconventional conflict, this is an unlikely assumption and the nature and quality of the intelligence operations can be considered one of the central elements of the conflict. The discussion of adjudication methodologies below addresses implementing intelligence operations, both in gathering data that would populate a set of perceptions and in influencing the opinions of others. ***This is a key part of the model design and is not included in the ontology***.

Creating the Metric Model

The metric classes and their instantiated objects contain the outputs of the simulation. However, a close reading of the ontology will reveal that they are not connected! This was done on purpose. Saying that the value of one metric affects the value of another metric is a model. Describing how one value affects the other is an elaboration of the model. This metric model is not properly part of a general ontology, but part of the model of a specific situation (even if it later becomes evident that the metric model is generalizable) and requires VV&A.

This metric model differs from the metric models of the GTO sets, shown in Fig. 5.38, Page 138. Those metric models represent the beliefs of the GTO set owners, not necessarily the true inferential relations among the metrics. This metric model is meant to represent the true relations, based on theories.

Figure 7.6 shows a partial metric model, taken from ISSM (Hartley, 2006c).

- The action, "Rebuild Roads," yields a measure of performance metric, *RebuildRoadsMOP*.
- The value of this metric affects three rating metrics (after suitable delays), *InterventionInvestmentRating*, *InterventionJobsRating*, and *Intervention-TransportationRating*.
- The *InterventionTransportationRating* is also affected by other metrics and its value affects *EfficientMarketsRating*, *CriticalIndustriesRating*, and other metrics.

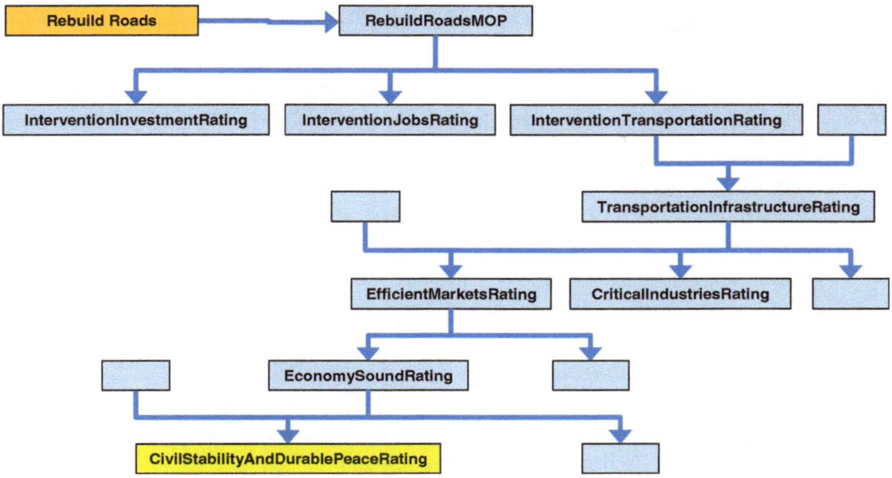

Fig. 7.6 Example of a partial metric model

- The *EfficientMarketsRating* is also affected by other metrics and its value affects the value of *EconomySoundRating* and other metrics.
- The value of *EconomySoundRating* affects *CivilStabilityAndDurable-PeaceRating* and other metrics.
- The final metric, which measures overall success (from the U.S. point of view), *CivilStabilityAndDurablePeaceRating*, is also affected by other metrics.

The import of this model is that, in a real-world situation, you can infer the value of the highest-level metrics, measures of political effectiveness (MoPEs), from the values of more accessible metrics. And you can do the same thing in a model where the calculated values of the "accessible" metrics can be based on theories with relatively high validity. The MoPEs are the metrics that represent answers to the questions the model is designed to answer.

Creating a metric model requires subject matter experts in several fields. It is not technically difficult because the ontology provides all the elements that need to be connected. However, it is very important because this is where the assertion is made that a certain state of the situation is better or worse than another state. Each connection requires a call to a specific theory or set of theories as justification. ***This is a key part of the model design***. That means that each metric represents an attachment point of one or more theories, ***which are outside of the ontology***.

This point deserves emphasis. Unless the model output consists only of a collection of numbers, requiring human analysis to establish the meaning of the model's results, ***the metric model provides the most important output of the model***. It provides the answer to "who won?" and similar questions.

Scenario Construction Aids

An important part of model construction occurs with the definition of a scenario and the work of modeling that scenario with the simulation. A totally fictitious scenario can be used in the initial construction and testing phases and has the advantage of controlling the features of the model that will be exercised. However, discussing a scenario taken from the real world exposes some important activities and vignettes from the real world can provide useful test cases and use cases.

First, the real world is so rich in detail that the simulation would bog down if it were all included. This is where modeling the scenario comes in. In Fig. 5.36, page XX, 12 sides were pictured. In that situation, many more sides could have been included. For example, all of the NGOs were represented by two sides by combining several into each. This is a modeling decision, based on a judgment that the extra complexity of additional sides would not yield significant differences of results. Second, in a fictitious scenario, data can be invented. In a real-world scenario, data have to be discovered. Despite the richness of the real world, data are hard to come by. Often the data that can be found must be used to infer the data that are needed. This is discussed in more detail in the Creating Data subsection of the "Gathering the Data" section below.

Figure 7.7 sketches the scenario definition process. The first step is to create a verbal description of the scenario.

The GTO sets provide initial assistance in modeling the scenario (Fig. 5.37, page XX). The distinct sides (owners) and their agendas (goals and tasks) will drive the simulation. Defining the GTO sets is an undertaking that is different in kind from that of creating the building blocks. Here the task is to get into the minds of the owners to discover what they want and what they think they should do to get what they want. In the ontology work, this required brainstorming by a group of subject matter experts (SMEs).

These GTO sets then must be converted into classes and subclasses in the model and connected to instantiations of the actors, actions and metrics that are specified in the GTO sets. Once the GTO sets have been defined, the owner rules for each GTO set can be defined. These first steps are illustrated in the left side of Fig. 7.7.

Even with the building blocks constructed and the scenario defined, as described thus far, and everything instantiated with the proper data, the simulation would not do anything. It needs a flow of events. This too is part of the scenario. The first part of the flow of events is the "road to war." (The second part of the flow of events will be described below in the Tactical Decision Rules section.) This starts out with a narrative of what has gone before—what leads up to the start point of the simulation. In a real-world scenario, this is obtained by observation. In a fictitious scenario, the concept must be created to determine what the initial values of the simulation will be. This narrative is needed to ensure that the initial values of the simulation are self-consistent.

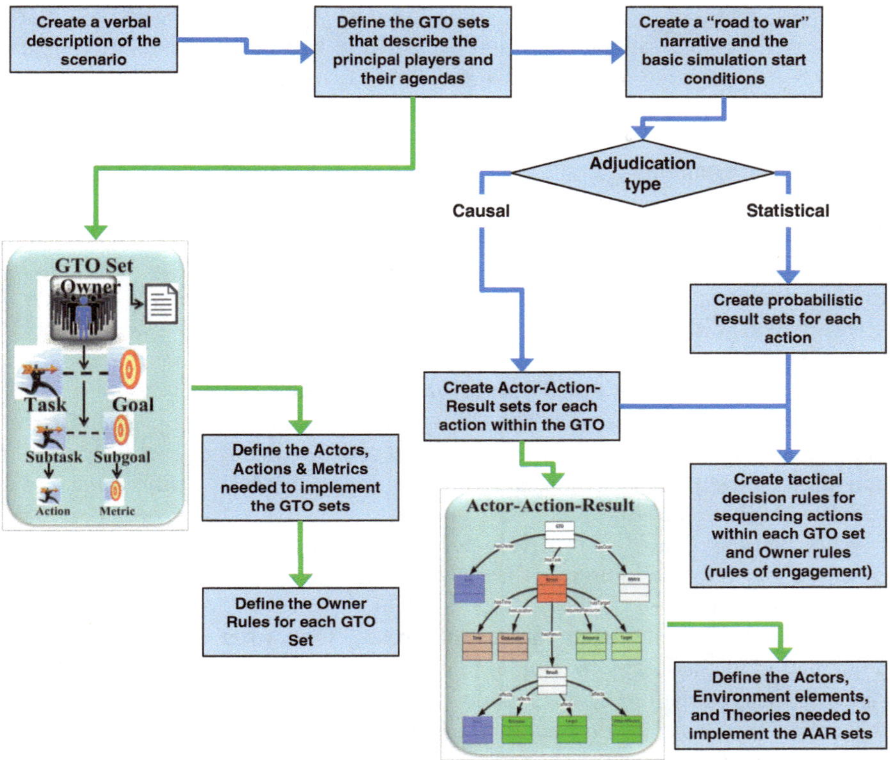

Fig. 7.7 Scenario definition process

Adjudication Methodologies

The second part of the scenario definition is adjudication, illustrated in the lower right section of Fig. 7.7. It should be noted that GTO sets also do not have any implied adjudication. That is, each owner has a set of goals, subgoals, and associated metrics for determining the status of the goals and each owner has a set of tasks, subtasks, and associated actions that the owner believes will accomplish the goals. There is no guarantee that the tasks, even if undertaken in an environment with no opposition, will lead to satisfaction of the goals! There are three basic adjudication alternatives: causal, statistical, and data-driven.

Causal Adjudication: There is a structure in the situation-dependent ontology that can be useful for either basic alternative. That structure is the AAR structure (Fig. 5.40, page XX). As noted in the discussion of the Action Ontology, there are 385 action classes. That means that 385 AAR structures must be examined. Probably some of the actions are similar enough to one another that some reduction in that number is possible. Each AAR structure defines a large number of variable environmental conditions, meaning that the work involved is daunting. However, this is not a problem caused by the ontology, but by the nature of the domain. Each of the AAR sets will define additional actors and will define the environmental elements that must be included. This alternative is illustrated in the lower-central section of Fig. 7.7.

This methodology has the advantage of an explicit call to some theory or theories to generate the central "result" in the model. This call isolates the use of theories, particularly social sciences theories, which is a great advantage in VV&A. The impact of validity questions concerning the theoretical bases of the overall model is clearly exposed. However, causal adjudication will result in a slower-running model.

Statistical Adjudication: The statistical adjudication has several sub-alternatives. One possibility is gathering a panel of SMEs and having them examine each action in the GTO sets. The panel would have to define a set of possible results for each action. Then, the panel would need to set probabilities of occurrence for each result. These probabilities could be indexed by environmental conditions prevailing at the time of the action. This alternative is illustrated in the center-right section of Fig. 7.7.

This alternative yields a relatively fast-running model, with validity dependent on the details of the definitions of the sets of results and the methodology chosen for assigning probabilities.

Combined Causal/Statistical Adjudication: The Army has used a mixed adjudication methodology in combat modeling. A fine-grained model is built and run with many variations of the situation, producing sets of probabilities of results that are parameterized by the conditions. In the main model, whenever adjudication is called for, the current conditions are matched against the set of fine-grained model runs and those results are plugged into the main model. In this case, the fine-grained models would be defined by the AAR structures.

However, without a well-defined organizing principle for the sets of conditions that must be addressed, both statistical adjudication alternatives risk missing needed situations and missing potential results.

Data-Driven Adjudication: The third type of adjudication relies on input data to determine the result. This allows for the direct investigation of the impact of different results.

A more complex variant of this type of adjudication can be seen in the problem of intelligence operations. Lt. General Flynn's description of the operations in Iraq clearly showed that the overall impact of intelligence operations can range from essentially zero to extremely high (Flynn & Ledeen, 2016). And from his description, the factors that change this impact may not be readily modeled. The ontology contains the essential elements (information operations, popular opinions, information collection, etc.) needed to describe the processes; however, the critical link between the actions and the results depends on these possibly un-modelable factors. It is possible that a set of input values that drive the results may be the best choice for this problem. Naturally, as with choosing theories, choosing such data values must be examined carefully during VV&A.

Tactical Decision Rules

The third part of the scenario is a narrative sketch of the actions to be taken at the beginning of the simulation and a means of computing what subsequent actions will be taken. (This is the beginning of the second part of the flow of events.) For example, the GTO sets do not have any priorities assigned to the tasks, subtasks or actions. In the real world, no side has the resources to perform all of the actions in its agenda at once. The owner and its subordinates prioritize the tasks and assign them to actors to perform them, based on the situation. In the simulations, this function can be represented by a set of tactical decision rules that has to be created for each owner. The tactical decision rules will be based on the state of the situation, as understood by the owner, the owner rules in force at the time, and the beliefs of the owner as represented by his metric model. These tactical decision rules will also be used in determining how actor–actor relationships change (e.g., are two factions still allies?) and how actor–thing relations change (e.g., who controls this city now?) as the situation changes.

There is no modeling law that requires that the tactical decision rules be maintained as a separate module; however, because of their data-like character and the need to support intense scrutiny during the V&V process, segregation may be advisable.

Culling the Simulation

It is conceivable that a major unconventional conflict could use every element and metric in the ontology; however, many conflicts will use a subset of them and some only a minor subset. The question is what is in the subset? The parts of the ontology provide much of what is needed to identify the parts that are not needed and to identify implied parts that are needed to ensure that the model is consistent in level of resolution and complete. Figure 7.8 represents the process of culling the unneeded parts of the simulation.

Reducing the Simulation

The first (horizontal) block of the figure represents the element and metric ontology categories. The subcategories are omitted (the "..." in the figure) and the second block represents the thousand+ element and metric classes. In that "major conflict" all of these will be needed. However, when fewer are needed, the first step in the process is to define the GTO sets, with their owners, actions, and metrics. Remove everything else. For a particular model, some of the Tasks (with their goals, subtasks, subgoals, actions, and metrics) may not be needed. Remove these. Further,

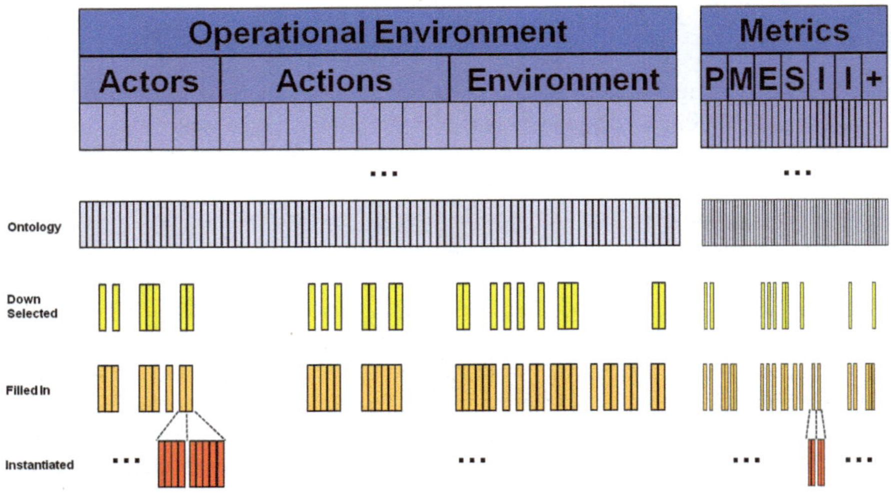

Fig. 7.8 Culling the simulation

some of the remaining subtasks (with their subgoals, actions, and metrics) may not be needed. Remove these. The third "Down Selected" block represents this step.

Filling in the Simulation

The fourth step consists of finding the elements and their metrics that are implied by the down-selected remnant. The stocks-and-flows relations will be helpful in finding elements that are related to the remnant elements that may needed to be added back. The semantic thesaurus will be helpful in identifying metrics (and their associated elements) that may also be needed.

The fifth step in the figure represents the process of instantiating one or more objects for each of the "filled in" elements and metrics.

The final result is a minimal model that will simulate the desired scenario, while maintaining completeness and consistency of resolution of the overall model.

Gathering the Data

No model runs without data. Part of the beauty of building a model from an ontology is the clear definition of the data needs that the ontology provides. At this point, all data are situation-dependent. Still, there are differences in the data needs for the situation-independent parts of the ontology and the situation-dependent parts. The situation-independent data are defined by the properties and the situation-dependent data are defined by the GTO sets, Rules sets, AAR sets, Relations sets, and Theory requirements.

Properties

Properties define the data needs for the ontology elements. Each element class has its own metrics and assigned set of properties. Thus, not all of the following properties will apply to any given element class. The instantiations of a particular element class will be defined by the individual property values for the instantiated entity with its instantiated metric(s).

Property Data: Entity Type

These data pertain to entities. There are three categories: primitive, activity, and location.

Primitive properties are those that are basic to the entity:

- **entityIdentity:** Name of entity; for an action, this is the identity of entity impacted;
- **entityQuantity:** Number of entities; e.g., if the name is "Truck, serial no. xxx" then the number is 1; if the name is "truck" the number could be 4000; for actions, this is the number of things acted on (produced, changed, added, subtracted, etc.);
- **entityMembersQuantity:** Number of members of the entity—applies only to things with members, like groups (only actors);
- **entityAnimate:** Indicator as to animate or inanimate status of entity (only actors);
- **entityMovable:** Indicator as to whether entity can be moved or not;
- **entitySpeed:** Entity's speed of movement (only actors);
- **entityWeaponry:** Entity's weaponry quantities (in general, actors or tangible things might have weapons); and
- **entityCanPerformAction:** Entity can perform an Action (yes or no, only actors).

Activity properties are those that describe a particular activity of an actor-type entity:

- **activityCoverage:** Entity activity in terms of coverage;
- **activityIntensity:** Entity activity in terms of intensity; and
- **activityQuantity:** Entity activity in terms of numbers of activities.

Location properties are those that describe the current location of an entity:

- **location:** Entity location is defined by one or more points, along a network, as an area, or by a density function over an area; and
- **locationDepthElevation:** Entity location depth or elevation.

Property Data: Relations Type

These data pertain to relations of entities. There are two categories: affiliation and authority.

According to Lt. General Michael Flynn, affiliation was critical information for Iraq (Flynn & Ledeen, 2016). Flynn said that the local people would wait until they could figure out who was going to win. Once they decided who was going to win, they would support that side and that was the side that received all the support and intelligence that was, in fact, critical to winning. Affiliation properties describe the current affiliations of an entity:

- **affiliationIdentity:** Name of thing with which entity is affiliated; for an actor this is actor's organization or parent organization; for a non-actor entity this is the entity's owner; for an action this is action's originator;
- **affiliationAuthorityIdentity:** Name of authority holder; for an actor this is the actor's superior; for environment this is the owner; for an action this is the action originator's superior; and
- **affiliationIntensity:** Intensity with which entity holds the affiliation or an entity's members hold the affiliation.

Authority properties describe the current authority of an actor-type entity:

- **authorityLevel:** Entity's authority level; and
- **authorityDistribution:** Type distribution of authority.

Property Data: Characterization Type

These data pertain to characterizations of entities. There are two categories: fixed characterization and other characterization.

Fixed characterizations include a set of predetermined characterizations of an entity:

- **levelAvailability:** Numeric or categorical level of availability of entity;
- **levelEffectiveness:** Numeric or categorical level of effectiveness of entity;
- **levelEfficiency:** Numeric or categorical level of efficiency of entity;
- **levelFairness:** Numeric or categorical level of fairness of entity;

- **levelHealthOrStrength:** Numeric or categorical level of health or strength of entity; and
- **levelInfluence:** Numeric or categorical level of influence of entity.

Other characterizations support the addition of other characterizations of an entity:

- **characterizationIdentity:** Name of some other characteristic of entity; and
- **levelCharacterization:** Numeric or categorical level of characterizationIdentity of entity.

Property Data: Environment Type

These data pertain to environmental entities. There are three categories: capacity, decision-making, and disaster or condition.

Capacity properties define the current capacity of an environment-type entity:

- **capacityFlowrate:** Capacity of entity in terms of flowrates such as number or volume per day; and
- **capacityQuantity:** Capacity of entity in terms of quantities such as weight, height, or volume.

Decision-making properties describe the decision-making of an element in the conceptual environment:

- **decisionMakingProcess:** Description of the decision-making process; and
- **decisionMakingQuality:** Numerical level of the quality of the decision-making.

Disaster or condition properties describe disaster or condition nature of elements in the natural environment:

- **indicatorDisasterOrCondition:** Indicator as to whether entity is disaster or condition;
- **canBeCausedByMan:** Indicator as to whether can be caused by man or not; and
- **canBeCausedByNature:** Indicator as to whether can be caused by nature or not.

Property Data: Result Type

These data pertain to actions. There are two categories: change and time.

Change properties describe the changes that comprise a particular result:

- **changeLevel:** Change to entity in terms of percent of possible change or in absolute numerical or categorical terms (only actions);
- **damageSeverity:** Damage to entity as a numeric or categorical severity level (only actions in damage and antiperson);

- **damageQuantity:** Damage to entity in numerical terms (only actions in damage and antiperson); and
- **damageCoverage:** Damage to entity in terms of the coverage over all components (only actions in damage and antiperson).

Time properties describe the times associated with a particular action:

- **timeDateTime:** Event occurrence as point in time;
- **timeDuration:** Event duration; and
- **timeFrequency:** Events per time period.

Situation-Dependent: GTO Data

Table 5.1, page XX, and Table 5.2, page XX, show the types of data that are required in defining GTO sets. For U.S. armed forces, the information for tasks, subtasks and goals can be derived from Army Field manuals under lines of effort (HQ DA, 2008b Oct) and (HQ DA, 2008a Feb). For other GTO sets the SMEs must use what published material can be found and their tacit knowledge of the real-world owners. The connections to the actions and metrics must be made using the definitions of these classes and the subtask and subgoal content. This is where the owner's metric models are created.

In 2016, Sebastian Gorka wrote a book about defeating the Jihadist movement (including ISIS, Al Qaeda, and others). He included a thorough discussion of the history, thought processes, and agendas of the significant groups and the movement as a whole. He also included a brief discussion of the appropriate agenda for the U.S., including three strategic tasks: (1) Deploy the truth (explain the true nature of the enemy in a massive counterpropaganda campaign; strengthen the counterpropaganda efforts of our allies and partners); (2) Help others fight their own wars (employ foreign internal defense operations; secondarily, use our forces to locate and kill the worst of the worst in the jihadist movement); (3) Win the war at home (educate and train the armed services, the federal, state, and local police forces, and the intelligence community in the enemy threat doctrine; employ the - New York Police Department's model of counterterrorism intelligence across the federal government) (Gorka, 2016).

Situation-Dependent: Rules Data

Owner rules are not just situation-dependent in the sense of the larger situation of a particular scenario, but will generally depend on the situation at a particular simulation time and place. The available rules can be determined independent of the situation, e.g., "use force only if directly threatened," "use force only on enemy combatants, but at any time they may be engaged," "use force on any target at any time, including non-combatants, combatants, and infrastructure of any type." However, assigning the rules to a particular owner will require subject matter experts to determine the past (or present) behaviors of forces of a given type to infer which rules would be selected by the owner.

It should be expected that the results of a simulation will be very sensitive to the rule definitions and the choices made in assigning the rules for use. There are frequently discussions in the media concerning the appropriateness of the rules of engagement used in actual situations, particularly when things are not going well. Some of these rules may be considered as scenario-defining inputs to allow for direct examination of the impact of changing rules.

Tactical decision rules are similarly extremely situation-dependent. Because they are essentially rules for deciding what to do next in a given situation, they are critical drivers of the simulation. The rules will depend on the state of the situation as understood by the owner, the owner rules in effect at the time, and the owner's beliefs as represented by his metric model (Fig. 5.38, page XX). As such, tactical decision rules deserve creation and review by subject matter experts. Generally, these rules will look like computer code, rather than data; however, they should be segregated in the model to aid in the intense scrutiny they should receive during the V&V process.

Situation-Dependent: AAR Data

Subject matter experts will be required to create the details of each AAR set. More than that, SMEs will be required to define which AAR sets need to be created. This latter function is at least helped by the centrality of a particular action or closely linked group of actions for each AAR. That means that while large, the number of potential AARs is finite.

Situation-Dependent: Relations Data

As with the owner rules, the relations data can be developed independently of the particular situation and their assignment will depend on the situation. The tactical decision rules are the appropriate means for making these assignments after the initiation of the simulation.

Situation-Dependent: Theory Data

As described above, a model built from this ontology provides the clear pointers to the places where theories are needed. Wherever the model updates the value of a metric, an invocation of some theory is required to establish the new value (Fig. 5.29, page XX). Wherever the model uses an AAR structure (or some analog) to define the result of an action, a call to some theory is required to establish this result (Fig. 5.40, page XX). The creation of the central metric model requires calls to theories for each inferential connection among metrics (Fig. 7.6, page XX).

A pdf document containing a number of theories by discipline can be found at http://drdeanhartley.com/HartleyConsulting/VVATool/VVA.htm. Page down to the figure with the title of "Static Data Entry Details" that also has the label "Validating the Conceptual Model" at the bottom of the figure and click on the download link below the figure. Each theory has a preliminary validity metric assigned to it.

Data Sources

Marina Arbetman-Rabinowitz provides a very useful description of the problems with social data and what needs to be done (Arbetman-Rabinowitz, 2010).

> Social scientists use and produce massive amounts of information in different formats from census [sic], opinion polls, surveys and experts, which can be qualitative or quantitative. These data get collected with different periodicity, sometimes on a monthly or annual basis, time series or with no continuity. This approach offers little value beyond that project's specific use. In other cases, data sets that had been collected for many years stop being collected. Often the data gets buried and forgotten, and a massive effort gets lost. Even more common data collection efforts constitute a one-time endeavor and as they stop they are not suitable to make inferences beyond that point.
>
> The situation is suboptimal at the moment. One of the many challenges is to create a good data base for assessment, with capabilities to compare pooled time series, capacity building for data updates, and an effective dissemination process for researchers. The data is collected by research institution, government statistical offices, NGOs, and special project teams and then archived without any plan for dissemination. Social scientists, journalists, marketing experts, all of them miss a great opportunity to advance knowledge. Part of the problem is that data has become a public good. Everybody wants to use it and nobody wants to pay for it, or pay to maintain archives or update it. Some issues with dissemination are technical and administrative and sometimes the separation between the metadata and the statistical data. The reasons are many; but, the question is then how do we deal with the issue of quality and dissemination in a world where the Internet has "raised the bar" in terms of quality and reduced the cost?

She also provides a large set of data sources and notes about the data they contain. Some potential sources are listed in the following tables from her paper. Her preface to the tables says, "but keep in mind that the sources keep changing leaving the researcher with incomplete and outdated information. In all, the number of possible data sources is immense. Most data sets encounter grave problems of quality, periodicity and/or timeliness" (Tables 7.1, 7.2, 7.3, and 7.4).

There are also specialized data sets on terrorism (Table 7.5):

Alok Chaturvedi also provided a sample set of resources used in SEAS modeling (Tables 7.6, 7.7, and 7.8).

Table 7.1 Conflict data: sources

Correlates of War Project: 1816–2006
State System membership, militarized disputes, formal alliances, national material capabilities. Not updated. Some data sets are historical, for example: Militarized inter and intra state disputes 1816–2006, Includes national contiguity 1816–2006
Available at: http://www.correlatesofwar.org/
UCDP/PRIO Armed Conflict v.4-2007, 1946–2006
Conflict-year data within countries where at least one party is the government
Available at: http://www.pcr.uu.se/research/UCDP/data_and_publications/datasets.htm
UCDP Armed Conflict Dyadic v.4-2006, 1989–2005
Dyadic version of the UCDP/PRIO armed conflict data
Available at: http://www.pcr.uu.se/research/UCDP/data_and_publications/datasets.htm
UCDP Conflict Termination dataset v.2.0, 1946–2006
Conflict level and conflict year data
Available at: http://www.pcr.uu.se/research/UCDP/data_and_publications/datasets.htm
UCDP Peace Agreement Dataset v. 1.0, 1989–2005
Data set that includes peace agreements signed by at least two warring parties
Available at: http://www.pcr.uu.se/research/UCDP/data_and_publications/datasets.htm
UCDP Battle-Deaths Dataset v.4.1, 2002–2005
Conflict level and conflict year data on number of deaths in internal conflicts
Available at: http://www.pcr.uu.se/research/UCDP/data_and_publications/datasets.htm
UCDP Non-State Conflict Dataset v.1.1, 2002–2005
Data set on internal conflicts where neither warring party is the government
Available at: http://www.pcr.uu.se/research/UCDP/data_and_publications/datasets.htm
UCDP One-Sided Violence Dataset v.1.2 1989–2005
Intentional attacks on civilians by the government or third party
Available at: http://www.pcr.uu.se/research/UCDP/data_and_publications/datasets.htm
Managing Intrastate Low-intensity Conflict (MILC) v. 1.0 1993–2004
Third party actions in Interstate conflict
Available at: http://www.pcr.uu.se/research/UCDP/data_and_publications/datasets.htm

In 2010, the following provided a number of sources on the respective topics that may be useful:

- Abdollahian, Kugler, Nicholson, and Oh discussed politics and power (Abdollahian, Kugler, Nicholson, & Oh, 2010);
- Alan Graham discussed economics and markets (Graham, 2010);
- William Bennett discussed the media and influence (Bennett, 2010);
- Corey Lofdahl discussed governance and society (Lofdahl, 2010);
- Bhavnani, Miodownik, and Riolo discussed groups and violence (Bhavnani, Miodownik, & Riolo, 2010);
- Kott and Skarin discussed insurgency and security (Kott & Skarin, 2010); and
- Duong, Turner, and Selke discussed crime and corruption (Duong, Turner, & Selke, 2010).

Table 7.2 Politics and governance data: sources

Transparency International Corruption Perception Index 1995–2007
Transparency across revenue, national accounts, perceived corruption
Available at: http://www.transparency.org/policy_research/surveys_indices/cpi
Bank's Cross National Time Series Data Archive 1815–2006
Demographic, Economic, Development, Stability, Regime Duration data
Available at: http://databanksinternational.com/
Freedom House 1972–2007
Civil, Political and Economic freedom
Available at: http://www.freedomhouse.org/template.cfm?page=15
Human Development Reports 2005–2008
Basic needs and quality of life data (UNDP)
Available at: http://hdr.undp.org/en/statistics/
World Values Survey, 1981–2004
Surveys conducted in waves, individual perceptions of regime satisfaction, economic satisfaction, well-being, general values
Available at: http://www.worldvaluessurvey.org/
The Urban Governance Initiative Report Cards
World Bank—perception of institutional reliability, corruption, urban and rural development. Available at: http://web.worldbank.org/WBSITE/EXTERNAL/WBI/EXTWBIGOVANTCOR/0,,contentMDK:20746471~pagePK:64168445~piPK:64168309~theSitePK:1740530,00.html
World Freedom Atlas 1990–2006
The geo-world atlas is a geo-visualization tool for world statistics
Available at: http://freedom.indiemaps.com/

Over several years, I have compiled and organized a large number of data sources by data category and subcategory. Each source has a name and rating and a text description. The value of these data sources is less certain than the sources described above, as mine were not compiled by social scientists; however, they may prove to be useful. A pdf document containing a large number of data sources by data category can be found at http://drdeanhartley.com/HartleyConsulting/VVATool/VVA.htm. Page down to the figure with the title of "Static Data Entry Details" that also has the label "Verifying and Validating the Data" at the bottom of the figure and click on the download link below the figure.

Creating Data

Some data are simply not available. Situations in which the existing data must be interpolated, reformatted, or otherwise massaged are frequently found in combat models and will be prevalent in unconventional conflict models. Other situations may require creating potential data values using subject matter experts and running models with variations of these potential values to determine the impact of these variations. These examples generally apply to historical data; however, current data pose a different problem.

Table 7.3 National economic/socio-demographic data: sources

Central Bank Websites
URL for Central Banks of most countries in the world
Available at: http://www.bis.org/cbanks.htm
Earth Trends
A compilation of a variety of social. Economic and environmental data
http://earthtrends.wri.org/searchable_db/index.php?step=countries&cID[]=6&theme=5&variable_ID=353&action=select_years
Government Financial Statistics
Available at: www.imf.org/external/pubs/ft/gfs/manual/gfs.htm
International Financial Statistics
Available at: http://www.imfstatistics.org/imf/logon.aspx
International National Statistics 1950–1993
Available at: *International National Statistics, UK, MacMillan Reference, LTD. 1998 [Europe; Africa, Asia & Oceana; The Americas, 1950–1993*
Penn World Tables
Provides purchasing power parity and national income accounts converted to international prices for 188 countries for some or all of the years 1950–2004
Available at: www.imf.org/external/pubs/ft/gfs/manual/gfs.htm
World Development Indicators
Available at: www.worldbank.org/data/wdi2004/
UN Statistical Data bases, Yearbooks and Reports
Available at: http://unstats.un.org/unsd/databases.htm

The data needed to describe a current situation will not be found in a book or database because the data have not yet been collected, analyzed, and posted. If someone else is tasked with performing these functions, the modeler will hope to have some influence on these processes, defining what is needed and why, etc. However, this influence is likely to be minimal, leading to similar data creation methods as those applied to historical data.

Active data collection in an unconventional conflict can be problematic or downright dangerous, depending on the situation and the type of data being collected. Further, the process of designing, administering, and analyzing questionnaires, polling, etc., represent separate disciplines and should probably be left to experts.

However, there is a type of passive data collection that can lead to the creation of needed data on current situations that is accessible and relevant to modeling unconventional conflict. This creation methodology was referred to earlier as text-parsing. Essentially, electronic versions of text documents or electronic communications are interpreted and analyzed by computer programs, yielding various types of data.

Yohai, Riordan, Duchon, and Paley discuss a methodology for identifying key entities and understanding their relationships through topics (Yohai, Riordan, Duchon, & Paley, 2013b). This methodology extracts "novel entity characteristics

Table 7.4 Region- or country-specific data: sources

Europa World Yearbook
Available at: www.europaworld.com/
London: England, Europa Publications Limited
Africa South of the Sahara
Available at: http://www-sul.stanford.edu/depts/ssrg/africa/www.unesco.org/unesdi/index.php/eng/repertoire/tous
Europa Publication (London), 1971–2000
"African Statistical Yearbook [ALSO African National Statistics, ASN=ASY]"
https://unp.un.org/details.aspx?pid=15447
The Far East and Australasia 1970–2000
Reference Library Book
London, Europa Publications, 1970–2000
The Middle East and North Africa 1948–1959, 1974–2000
Reference Library Book
London, Europa Publications, 1974–2000
Statistical Yearbook for Asia and the Far East 1968–1969
Reference Library Book
United Nations
Statistical Yearbook for Asia and the Pacific, United Nations, Bangkok, Thailand, Economic and Social Commission for Asia and the Pacific, 1974–1984, 1986–1990, 1992–1999
Reference Library Book
Statistical Yearbook for Latin America, Santiago, Chile, United Nations, Economic Commission for Latin America, 1975–1984
Reference Library Book
Statistical Yearbook for Latin America and The Caribbean, Santiago, Chile, Economic Commission for Latin America and the Caribbean, 1986–1999
Reference Library Book
The Statistical Abstract of Latin America CEPAL
Reference Library Book
International Historical Statistics Mitchell
Reference Library Books

Table 7.5 Terrorism data: sources

International Terrorism: Attributes of Terrorist Events (ITERATE)
The ITERATE project quantifies data on the characteristics of transnational terrorist groups, their activities that carry international impact and the environment in which they operate
Available at: http://library.duke.edu/data/collections/iterate
Global Terrorism Database
The GTD is an open-source database including information on terrorist events around the world from 1970 through 2014 (with annual updates planned for the future). Unlike many other event databases, the GTD includes systematic data on domestic as well as international terrorist incidents that have occurred during this time period and now includes more than 140,000 cases
Available at: https://www.start.umd.edu/gtd/

Table 7.6 Economic data: sources

World Development Indicators 2007, World Bank, http://web.worldbank.org/WBSITE/EXTERNAL/DATASTATISTICS/0,,contentMDK:21298138~pagePK:64133150~piPK:64133175~theSitePK:239419,00.html
CIA World Factbook, https://www.cia.gov/library/publications/the-world-factbook/
World Bank, "Afghanistan: Data, Projects & Research," http://web.worldbank.org/WBSITE/EXTERNAL/COUNTRIES/SOUTHASIAEXT/AFGHANISTANEXTN/0,,menuPK:305990~pagePK:141159~piPK:141110~theSitePK:305985,00.html
IMF Country Report, http://www.imf.org/external/country/index.htm
United Nations Statistical Yearbook, http://unstats.un.org/unsd/pubs/gesgrid.asp?id=368
United Nations Office of Drugs and Crime, http://www.unodc.org/unodc/index.html
Energy Information Agency, U.S. Department of Energy, http://www.eia.doe.gov/emeu/cabs/contents.html
National Statistical Websites
• Mexico: http://www.inegi.gob.mx/inegi/default.aspx
• Bolivia: http://www.ine.gov.bo/
• The Philippines: http://www.bles.dole.gov.ph/
• Etc.

Table 7.7 Public opinion data: sources

WorldPublicOpinion.org, http://www.worldpublicopinion.org/
Pew Global Attitudes Project, Pew Research Center, http://pewglobal.org/
Reports by NGOs, *Afghanistan in 2006: A Survey of the Afghan People,* The Asia Foundation, http://www.asiafoundation.org/pdf/AG-survey06.pdf

Table 7.8 Security/political system data: sources

Freedom House rankings of political and civil liberties, http://www.freedomhouse.org/template.cfm?page=25&year=2007
The Minorities at Risk Project, University of Maryland, http://www.cidcm.umd.edu/mar/
Failed State Index, Foreign Policy, http://www.foreignpolicy.com/story/cms.php?story_id=3865
GlobalSecurity.org, http://www.globalsecurity.org/
International Crisis Group country reports, http://www.crisisgroup.org/home/index.cfm?id=1266&l=1
Congressional Research Service reports, http://www.fas.org/sgp/crs/row/index.html
Center for Strategic and International Studies, http://www.csis.org/index.php
The Brookings Institution, http://www.brook.edu/
Rand Corporation, http://www.rand.org/research_areas/international_affairs/
Strategic Studies Institute, U.S. Army War College, http://www.strategicstudiesinstitute.army.mil/
The Military Balance, The International Institute for Strategic Studies, http://www.iiss.org/publications/the-military-balance
The Middle East Military Balance, Jaffee Center for Strategic Studies, Tel Aviv University, http://www.tau.ac.il/jcss/balance/index.html
United States Institute of Peace reports, http://www.usip.org/pubs/index.html

and relationships from text." "Starting with known persons of interest, we can identify their key characteristics and understand the relationships that they share with other entities. Alternatively, starting with a topic of interest, the model outputs show the network of players associated with the topic, and the strength of that association."

Wedgwood, Ruvinsky and Siedlecki mention the exploitation of information found in news data, developed for the DARPA Integrated Crisis Early Warning System (ICEWS) (Wedgwood, Ruvinsky, & Siedlecki, 2013). While this paper focuses on other issues, they describe a system that uses text-parsing, among other techniques, to extract useful data from electronic versions of news reports.

Another part of the extraction of data from electronic sources is known as event coding, which is the "process of ingesting information from freeform text such as news stories, extracting information about things that happened (events), and coding these events in a standard way (Van Brackle & Haglich, 2013)." Van Brackle and Haglich discuss the structure of Jabari, their event coder and improvements to it.

VV&A of the Model

VV&A is an integral part of modeling. Frequently, it is thought of as an expensive add-on to the modeling process; however, any good modeler is checking and testing his or her model during the design and implementation process and endeavors to have the results accepted when the work is finished. This is the core of VV&A. A good modeler may even seek the services of independent, adversarial-minded tester whose stated mission is to break the model. Alternatively, both the modeler and tester may work for the same program manager, with neither reporting to the other. In this way, the tester plays the role of "brother heretic".

The ontological underpinnings of this modeling approach ensure that the problem of testing for coverage is relatively simple. The things that should be covered are explicitly modeled (or not modeled). The object-oriented implementation makes it relatively simple to verify the implementation of the conceptual model and perform validation on the more mundane aspects of the model. The explicit exposure of the use of theories (see the section on Situation-Dependent: Theory Data, above, for the points of attachment of theories) makes the assessment of the validity of the more difficult aspects of the model easier.

The advice in the final section of Chapter 6 should be applied:

- VV&A must be entrenched, with an explicit VV&A management process;
- System configuration management is required;
- V&V metrics for PMESII models are critical;
- The theories used in the system must be identified; and
- PMESII risk management and mitigation are required.

Chapter 8
Conclusion

This book is not the final word on modeling unconventional conflict. It represents a distillation of what is known and what has been done with regard to modeling unconventional conflict at the theater level up to this point. However, as has been made clear in the body of the book, there is much that needs to be learned, particularly regarding the social theories that must drive the estimates of outcomes of unconventional conflict activities.

Review

Chapter 2 presented what we know about unconventional conflict, providing descriptions of the operations that comprise unconventional conflicts and descriptions of analyses of the domain. Chapter 3 described a number of attempts at modeling all or parts of the unconventional conflict domain, describing modeling constructs that have been created in the attempt to get a handle on modeling the domain, describing technical modeling approaches that have been used, describing a fair number of models, and describing the results of workshops held to define the modeling requirements for the domain. Chapter 4 focused on the modeling construct that has proved to be the most durable and useful in organizing our modeling attempts in the domain, the DIME/PMESII paradigm.

Chapter 5 introduced the concept of an ontology as a means of describing a domain of knowledge and applying the ontology to unconventional conflict. This ontology was designed to contain the bulk of what we know about the domain. That means it had to support the information from Chapter 2, support the modeling needs explored in Chapter 3, and incorporate the DIME/PMESII paradigm. Such an ambitious undertaking required exploring the large body of narrative descriptions of unconventional conflict, understanding the many modeling attempts, and being willing to use ontological material from a number of sources, which were described in the chapter. The final sections of the chapter were devoted to describing the

D.S. Hartley III, *Unconventional Conflict*, Understanding Complex Systems,
DOI 10.1007/978-3-319-51935-7_8

structure of the ontology (also providing access to lists of some of the actual contents).

Chapter 6 was devoted to the critical subject of verification, validation, and accreditation (VV&A). Fortunately, by the time modeling unconventional conflict had become necessary, VV&A of other types of models had reached a useful level of maturity. Adapting what was known about "standard" VV&A was non-trivial, but possible because of this maturity.

Chapter 7 brought this all together, describing how the ontology supports building models that are complete (in the sense of including all of the necessary elements of unconventional conflict for a particular modeling purpose), parsimonious (in the sense of including no more than is necessary for the purpose), and consistent in level of resolution.

Recap of the Major Points

This book is structured to build to the exposition of Chapter 7, which describes a new method of constructing models of unconventional conflict. However, a number of significant points were made throughout the book.

About Unconventional Conflict

- Unconventional conflict is complex because of the number of relevant actors (with their inherent complexity and variety), the number of actions that can be undertaken, the importance of the passive environment, and the number of relevant metrics. This complexity means that understanding unconventional conflict is very difficult. For example, answering the simple question "who won?" is not simple for unconventional conflict. Unconventional conflicts often take place in unstable countries, so predicting instability is a major need and that is very hard to do.
- Because unconventional conflict is the norm now, understanding it is a necessity, despite the difficulties. Unconventional conflict is the union of OOTW and IW. It requires the explicit use of DIME/PMESII+ variables (includes some kinetic and environmental variables). It also requires HSCB, military, and some physical theories in explaining the results of actions.

DIME/PMESII Paradigm and Ontologies

- The creation of the DIME/PMESII paradigm was a major step forward. It called out the difference between the levers of national power (DIME) and the state

variables that described the situation (PMESII), which are the key to detecting changes in the situation.
- The creation of the IW Ontology was another major step forward in describing unconventional conflict and providing a tool for better modeling. It contains "everything" known about unconventional conflict in a holistic construct. It also separates situation-independent modeling from situation-dependent modeling. Finally, it isolates calls to theory to allow better validation efforts.

Modeling Unconventional Conflict

- Modeling conventional war and unconventional conflict have many similarities; however, there are many differences. The gaps in modeling unconventional conflict lie mostly in the social theory area. Also, there is a difference in implied requirements: rather than starting with combat forces and inferring support structure, unconventional conflict modeling starts with parts of what had been support structure and must infer the need for other support structure and for combat forces.
- There have been many models of conventional war at the theater level. Both conventional war models and unconventional conflict models must limit their levels of resolution in order to achieve reasonably short run times. However, identifying consistent lowest-level units across naval, air, and ground forces for conventional war is simpler than for unconventional conflict. For modeling unconventional conflict, the best lowest-level units are the important people, the important groups/organizations, and the important demographic groups; however, modeling unconventional conflict at the theater level does not include the "how the mind works" level. For conventional war, the time span will generally be measured in months; for unconventional conflict, the time span will often be measured in years.
- There is a base of work on modeling unconventional conflict, including the development of modeling constructs and the construction of models and systems of models. These efforts have produced useful results, both in analyses of actual conflicts and in producing information about what works and what does not work. Some of these models have been built specifically to address unconventional conflict and some have been the result of creating systems of existing models. Their starting points, in all cases, were based on initial, partial concepts of unconventional conflict and required modification as greater understanding was achieved.
- Ontological software design is easier and better because it starts with a holistic description of unconventional conflict that includes a detailed definition of the parts as intimately interconnected and explicable only by reference to the whole.

VV&A of Unconventional Conflict

- The definition of VV&A for unconventional conflict was another major step forward. The first commonality between conventional war models and models of unconventional conflict is that the VV&A process must be entrenched for both types of models. Entrenched VV&A includes all testing from model conception through model retirement. To be useful, entrenched VV&A requires maintaining records of all testing throughout the life of the model. This brings up the second commonality: strict version configuration control is required and must be attached to the testing results.
- The major difference between the VV&A processes for the two types of models lies in the metrics required for assessing validity. Unconventional conflict models rely on the use of social theories in determining the results of actions, whereas conflict models (at least those in the past) rely more heavily on physical theories, with stronger validity. For unconventional conflict, this requires both an emphasis on PMESII metrics and the understanding that unconventional conflict models will require much more care in their use and assessments of their results. While risk assessment and mitigation—and the understanding of residual risks—are common to both cases, the likelihood of larger risks is inherent in unconventional conflict models.

Concluding Thoughts

Chapter 7 presents a process for modeling unconventional conflict. However, before this process could be presented in its entirety, the supporting body of knowledge had to be explored. For this reason, there is a chapter that focuses on the description of unconventional conflict and the analyses that have been performed on it. There is a major chapter on modeling, with a concentration on past efforts at modeling unconventional conflict. There is a chapter on the precursors to the ontology and one on the ontology itself. And there is a chapter on the critical issue of VV&A.

Unconventional conflict is a complex, messy thing. It normally involves multiple actors, each with its own unique agenda and concept of legitimate actions. The unconventional conflict ontology supports holistic modeling. Holistic modeling means that we can both see the entire picture of what needs to be modeled and make informed decisions about what to model and what to omit. The unconventional conflict ontology also separates the things we understand best from the things we understand least. That is, it separates the easy (relatively speaking) things from the hard things and allows us to put more effort into deciding how to address the hard parts. These "hard parts" include adjudication (the choice of theories and the implementation of theories), the GTO sets (metric models), the tactical decision rules, and the modeling of intelligence operations (ground truth versus perception).

This separation also means that we can perform verification, validation, and accreditation more efficiently, and can describe the competence of the model more accurately.

If all of this seems complex and the resulting models seem problematic, then you have understood the exposition correctly. Even with the ontological approach to modeling unconventional conflict, the modeling process will be complex. And, even with the disentanglement of the "hard parts" from the easier parts, there is no guarantee that the hard parts will be modeled correctly.

It is important to realize that as long as unconventional conflicts take place, someone has to deal with them. Without the support of a coded model, that someone must rely on his or her mental model of the situation and the likely responses of other parties in making decisions. Just as Napoleon was regarded as a strategic genius in executing the conventional war of the time, there may be a genius in charge of conducting operations in an unconventional conflict. But if there are geniuses, there are many more "average" strategists and some awful strategists, with no guarantee of the quality of the one in charge for the next conflict. A coded model has the advantage over a mental model of being specified, so that if it is seen to be in error, that error can be identified, addressed, and mitigated, if not fully corrected. Further, in using a mental model, it is easy to forget to consider something. The coded model will, at least, consider everything that is specified—every time.

The core message of this book is that modeling unconventional conflict is now easier than it had been because of the development of an ontology that describes the unconventional conflict domain.

References

Abdollahian, M., Baranick, M., Efird, B., & Kugler, J. (2006). *Senturion: A predictive political simulation model*. Washington, DC: National Defense University.

Abdollahian, M., Kugler, J., Nicholson, B., & Oh, H. (2010). Politics and power. In A. Kott & G. Citrenbaum (Eds.), *Estimating impact* (pp. 69–79). New York: Springer.

Alberts, D. S., & Hayes, R. (1995). *Command arrangements for peace operations*. Washington, DC: National Defense University.

Allard, K. C. (2002). *Somalia operations: Lessons learned*. Oregon: University Press of the Pacific.

AMORS III. (1995). *Third Asia-Pacific Military Operations Research Symposium: Operations research for regional security issues in peace and low intensity conflict*. Bangkok, Thailand: AMORS.

Arbetman-Rabinowitz, M. (2010). Time to measure up: Dirty little secrets in data collection or: On how econometrics is easier without data. In A. Woodcock, M. Baranick, & A. Sciarretta (Eds.), *The human social culture behavior modeling workshop* (pp. 243–266). Washington, DC: National Defense University.

Asimov, I. (1951). *The foundation trilogy*. New York: Doubleday & Company.

Avruch, K., Narel, J. L., & Combelles-Siegel, P. (1999). *Information campaigns for peace operations*. Washington, DC: CCRP.

BAE Systems. (2007, February 22). *Battle command LIVE: Liquid information virtual environment*. Arlington, VA.

Bar, S. (2013). A toolbox for countering extremist religious doctrines. In D. M. Nicholson & D. D. Schmorrow (Eds.), *Advances in design for cross-cultural activities, Part II* (pp. 386–397). Boca Raton, FL: CRC Press.

Bell, S., Caglayan, A., Cassani, L., & Murdie, A. (2013). Predictive societal indicators of radicalism—Forecasting political violence. In D. M. Nicholson & D. D. Schmorrow (Eds.), *Advances in design for cross-cultural activities, Part II* (pp. 427–436). Boca Raton, FL: CRC Press.

Bennett, W. H. (2010). Media and influence. In A. Kott & G. Citrenbaum (Eds.), *Estimating impact* (pp. 135–178). New York: Springer.

Berg, S. T., & Hamby, L. B. (1995). *Regional stability assessment: Recognizing opportunities, identifying threats, and managing risk in emerging countries*. McLean, VA: Booz-Allen & Hamilton.

Berman, E., & Kuznar, L. A. (2010). Economic theories of conflict for COIN/STTR. In A. Woodcock, M. Baranick, & A. Sciarretta (Eds.), *The human social culture behavior modeling workshop* (pp. 30–33). Washington, DC: National Defense University.

D.S. Hartley III, *Unconventional Conflict*, Understanding Complex Systems,
DOI 10.1007/978-3-319-51935-7

Bhavnani, R., Miodownik, D., & Riolo, R. (2010). Groups and violence. In A. Kott & G. Citrenbaum (Eds.), *Estimating impact* (pp. 205–237). New York: Springer.

Bitinas, E. J. (2009). Pythagoras—An agent-based distillation system. In *Human Social Cultural Behavior (HSCB) Modeling Workshop II, 8–10 June 2009*. Washington, DC: National Defense University.

Blanco, R. J. (1994). *Counter-insurgency in Cuba: Why did Batista fail? AD-A284 697*. Ft Leavenworth, KS: US Army CGSC.

Bond, D., & Vogele, W. B. (1995). *Profiles of international "Hotspots"*. Cambridge, MA: Center for International Affairs, Harvard.

Box, G. E. (1979). Robustness in the strategy of scientific model building. In R. L. Launer & G. N. Wilkinson (Eds.), *Robustness in statistics* (pp. 201–236). New York: Academic.

Brundage, B., et al. (1998). *Analysis of U.S. involvement in multiple small scale contingencies—Failed state*. Washington, DC: OSD (PA&E).

Burke, M. C., & Self, E. C. (2008). *Population analysis: A methodology for understanding populations in COIN environments*. Monterey, CA: Naval Postgraduate School.

Burkle, F. M. (1995a). Complex, humanitarian emergencies: I. Concepts and participants. *Prehospital and Disaster Medicine, 10*(1), 36–42.

Burkle, F. M. (1995b). Complex, humanitarian emergencies: III. Measures of effectiveness. *Prehospital and Disaster Medicine, 10*(1), 48–56.

Burkle, F. M. (1996). Civil-military relations in complex humanitarian emergency response. *USPACOM Workshop: Modeling & Simulation Requirements for LRCs and Other Military Operations*. Monterey, CA.

Cababa, R. R. (1991). *Nation assistance—A misunderstood mission, AD-A240 519*. Newport, RI: Naval War College.

Carley, K. M. (2003). Dynamic network analysis. In R. Breiger, K. M. Carley, & P. Pattison (Eds.), *Dynamic network modeling and analysis: Workshop summary and papers* (pp. 133–145). Washington, DC: National Research Council.

Carley, K. M. (2004). Dynamic network analysis for counter terrorism. *MORS Global War on Terrorism (GWOT) Conference*. Washington, DC: MORS.

Carley, K. M., Morgan, G., Lanham, M., & Pfeffer, J. (2013). Multi-modeling and socio-cultural complexity: Reuse and validation. In D. M. Nicholson & D. D. Schmorrow (Eds.), *Advances in design for cross-cultural activities, Part II* (pp. 128–137). Boca Raton, FL: CRC Press.

Carment, D., Prest, S., St. Jean, L., Samy, Y., & Wyjad, K. (2010). Approaches to country risk analysis and early warning. In A. Woodcock, M. Baranick, & A. Sciarretta (Eds.), *The human social culture behavior modeling workshop* (pp. 33–64). Washington, DC: National Defense University.

Cassani, L., Caglayan, A., Alavedra, J., Das, S., Morgan, W., & Mooney, L. (2013). Population sentiment modeling in support of ISAF joint command. In D. M. Nicholson & D. D. Schmorrow (Eds.), *Advances in design for cross-cultural activities, Part II* (pp. 437–446). Boca Raton, FL: CRC Press.

Cassani, L., Caglayan, A., Morgan, W., Mooney, L., Boyles, A., & Weiss, B. (2013). Semantic integration of socio-cultural knowledge for complex operations. In D. M. Nicholson & D. D. Schmorrow (Eds.), *Advances in design for cross-cultural activities, Part II* (pp. 447–456). Boca Raton, FL: CRC Press.

Chairman of the Joint Chiefs of Staff. (1995). *Joint doctrine for military operations other than war, joint publication 3-07*. Washington, DC: Chairman of the Joint Chiefs of Staff.

Chairman of the Joint Chiefs of Staff. (2001). *Joint tactics, techniques, and procedures for joint special operations task force operations, joint publication 3-05.1*. Washington, DC: Chairman of the Joint Chiefs of Staff.

Chairman of the Joint Chiefs of Staff. (2011). *Joint operations, joint publication 3-0*. Washington, DC: Chairman of the Joint Chiefs of Staff.

Champoux, P., Trudel, M., & Thibault, G. (2004). *A lessons learned knowledge warehouse to support the army knowledge management command-centric*. Retrieved May 30, 2016, from DTIC.mil: www.dtic.mil/cgi-bin/GetTRDoc?AD=ADA428601

Chartuvedi, A. R. (2003). *SEAS-UV 2004: The theoretical foundation of a comprehensive PMESII/DIME agent-based synthetic environment*. Simulex.

Chaturvedi, A. (2009). Reference world: extensible toolkit for COCOMs. *Human, Social, Cultural Behavior (HSCB) Workshop II, June 8–10, 2009*. Washington, DC: NDU.

Cioppa, T. M., & Lucas, T. W. (2007). Efficient nearly orthogonal and space-filling Latin hypercubes. *Technometrics, 49*(1), 45–55.

Clair, C. D. (1993). *Humanitarian assistance and the elements of operational design, AD-293 291*. Ft Leavenworth, KS: US Army CGSC.

Clark, R. L. (1995). *Noncombatant evacuation operations: Major considerations for the operational commander, AD-A293 291*. Newport, RI: Naval War College.

Clausewitz, C. v. (1993). *On war*. New York: Knopf.

Cobb, L. (2010). The impact of social theory on model development. In A. Woodcock, M. Baranick, & A. Sciarretta (Eds.), *The human social culture behavior modeling workshop* (pp. 170–184). Washington, DC: National Defense University.

Combelles-Siegel, P. (1998). *Target Bosnia: Integrating information activities in peace operations*. Washington, DC: CCRP.

Cowan, J. W. (1995). *Operation provide comfort: Operational analysis for operations other than war, AD-A297 852*. Newport, RI: Naval War College.

Coyle, R. G., & Alexander, M. W. (1997). Two approaches to qualitative modelling of a Nation's drug trade. *System Dynamics Review, 13*(3), 205–222.

Coyle, R. G., & Millar, C. J. (1996). A methodology for understanding military complexity: The case of the Rhodesian counter-insurgency campaign. *Small Wars & Insurgencies, 7*(3), 360–378.

Davis, D. F. (1996). Peace operations analysis with bayesian belief networks. *13th International Symposium on Military Operational Research (ISMOR)*. England.

Davis, I. (2010). *RELATIONSHIP: A vocabulary for describing relationships between people*. Retrieved May 21, 2016, from vocab.org—A URI space for vocabularies: http://vocab.org/relationship/

Davis, P. K. (Ed.). (2011). *Dilemmas of intervention: Social science for stabilization and reconstruction*. Santa Monica, CA: RAND.

Davis, P. K. (2013). Fundamentals in empirical validation of and analysis with social science models. In D. M. Nicholson & D. D. Schmorrow (Eds.), *Advances in design for cross-cultural activities, Part II* (pp. 87–98). Boca Raton, FL: CRC Press.

Davis, P. K., & Cragin, K. (Eds.). (2009). *Social science for counterterrorism: Putting the pieces together*. Santa Monica, CA: RAND.

Davis, C., & Farnsworth, K. (1990). *After-action report of the Hurricane Hugo OFDA disaster relief team, PB91-210096*. Washington, DC: Agency for International Development.

Deckro, R. F. (2010). Issues in stability, security, transition, and reconstruction operations. In A. Woodcock, M. Baranick, & A. Sciarretta (Eds.), *The human social culture behavior modeling workshop* (pp. 381–401). Washington, DC: National Defense University.

Dell, R. F., & Sparling, S. J. (2008). Optimal distribution of resources for non-combatant evacuation. *76th Military Operations Research Society Symposium*. New London, CT: MORS.

Department of Defense. (2007, August 8). *Department of Defense Directive Number 5000.59*. Retrieved May 22, 2016, from DoD Issuances: http://www.dtic.mil/whs/directives/corres/pdf/500059p.pdf

Department of Defense. (2009, September 16). *dtic.mil, 3000.05*. Retrieved April 24, 2016, from DoD Issuances: http://www.dtic.mil/whs/directives/corres/pdf/300005p.pdf

Department of Defense. (2014, August 28). *dtic.mil, 3000.07*. Retrieved April 24, 2016, from DoD Issuances: http://www.dtic.mil/whs/directives/corres/pdf/300007p.pdf

Department of State. (2005). *Post-conflict reconstruction.* Retrieved May 4, 2016, from peacebuildingcentre.com: http://peacebuildingcentre.com/pbc_documents/US_State_Department_Post_Conflict_Essential_Tasks_2005.pdf

Department of State. (2016, April 19). *Updated foreign assistance standardized program structure and definitions*. Retrieved May 5, 2016, from Department of State: http://www.state.gov/f/releases/other/255986.htm

Ducote, B. M. (2001). *Challenging the a pplication of PMESII-PT in a complex environment, AY 2010-1*. Ft Leavenworth, KS: School of Advanced Military Studies, United States Army Command and General Staff College.

Duong, D., Turner, R., & Selke, K. (2010). Crime and corruption. In A. Kott & G. Citrenbaum (Eds.), *Estimating impact* (pp. 263–284). New York: Springer.

Durante, J., & Fox, W. P. (2015). Modeling violent acts in the Philippines. *Journal of Mathematical Sciences, 2*, 127–140.

Durch, W. J. (Ed.). (1993). *The evolution of UN peacekeeping—Case studies and comparative analysis*. New York: St. Martin's Press.

Dworken, J. T. (1993). *Military relations with humanitarian relief organizations: Observations from restore hope, CRM 93-140*. Alexandria, VA: Center for Naval Analyses.

Dworken, J. T. (1996). *Improving marine coordination with relief organizations in humanitarian assistance operations, CRM 95-161*. Alexandria, VA: Center for Naval Analyses.

Dziedzic, M., Sotirin, B., & Agoglia, J. (2008). *Measuring progress in conflict environments (MPICE)—A metrics framework for assessing conflict transformation and stabilization*. DTIC.

Efird, B. (2009). Senturion overview. In *COCOM requirements for human social and cultural behavior (HSCB) modeling, 8–10 June 2009*. Washington, DC: National Defense University.

Farnsworth, K., D'Silva, B., Lynch, L., Sullivan, E., Carroll, F., Little, M., et al. (1992). *Southern Africa drought assessment: March 24–April 29, 1992, PB93-237949*. Washington, DC: OFDA.

Fellman, P. V. (2015). The landscape of maritime piracy and the limits of statistical prediction. In P. V. Fellman, Y. Bar-Yam, & A. A. Minai (Eds.), *Conflict and complexity: Countering terrorism, insurgency, ethnic and regional violence* (pp. 199–220). New York: Springer.

Fellman, P. V., Bar-Yam, Y., & Minai, A. A. (Eds.). (2015). *Conflict and complexity: Countering terrorism, insurgency, ethnic and regional violence*. New York: Springer.

Ferleman, T., & Walters, D. (2010). The Middle East in 2040: A forecast of drivers of instability. In A. Woodcock, M. Baranick, & A. Sciarretta (Eds.), *The human social culture behavior modeling workshop* (pp. 116–135). Washington, DC: National Defense University.

Flynn, M. F. (1988, April). New York: William F. Battista. An introduction to psychohistory, Part I of II. In S. Schmidt (Ed.), *Analog Science Fiction/Science Fact, CVIII*(4) Davis Publications, Inc. (pp. 60–78).

Flynn, M. T., & Ledeen, M. (2016). *The field of fight*. New York: St. Martin's Press.

Geis, M. (1996). *Logistics and engineering requirements for humanitarian assistance operations, CRM 95-152*. Alexandria, VA: Center for Naval Analyses.

General Accounting Office. (1992). *United Nations: US participation in peacekeeping operations, GAO/NSIAD-92-247*. Washington, DC: GAO.

Goldstone, J. A. (2010). Principles for modeling to guide counterinsurgency operations. In A. Woodcock, M. Baranick, & A. Sciarretta (Eds.), *The human social culture behavior modeling workshop* (pp. 347–362). Washington, DC: National Defense University.

Gorka, S. (2016). *Defeating Jihad: The winnable war*. Washington, DC: Regnery Publishing.

Graham, A. K. (2010). Economics and markets. In A. Kott & G. Citrenbaum (Eds.), *Estimating impact* (pp. 91–134). New York: Springer.

Gregory, M., Engel, D., Bell, E., & McGrath, L. (2013). Identifying differences in cultural behavior in online groups. In D. M. Nicholson & D. D. Schmorrow (Eds.), *Advances in design for cross-cultural activities, Part II* (pp. 283–292). Boca Raton, FL: CRC Press.

Guermantes, L. (2010). Terrorism and counterterrorism M&S. In A. Woodcock, M. Baranick, & A. Sciarretta (Eds.), *The human social culture behavior modeling workshop* (p. 278). Washington, DC: National Defense University.

Gupta, D. K. (2010). Modeling the dynamics of terrorist movements: A macro interactive perspective. In A. Woodcock, M. Baranick, & A. Sciarretta (Eds.), *The human social culture behavior modeling workshop* (pp. 300–332). Washington, DC: National Defense University.

Hartley, D. S. (1975). *An examination of a distribution of TAC CONTENDER solutions, TM 101-75*. Washington, DC: NMCSSC.

Hartley, D. S. (1991). *Confirming the Lanchestrian linear-logarithmic model of attrition, K/DSRD-263/R2*. Oak Ridge, TN: Martin Marietta Energy Systems.

Hartley, D. S. (1995). Modeling psycho-social attributes in conflict. In AMORS (Ed.), *Third Asia-Pacific military operations research symposium: Operations research for regional security issues in peace and low intensity conflict*. Bangkok, Thailand: AMORS.

Hartley, D. S. (1996a). A framework for modeling psycho-social attributes in conflict. In A. Woodcock & D. Davis (Eds.), *Analytic approaches to the study of future conflict* (p. 300). Clementsport, Canada: The Canadian Peacekeeping Press.

Hartley, D. S. (1996b). *Operations other than war: Requirements for analysis tools research report, K/DSRD-2098*. Oak Ridge, TN: Lockheed Martin Energy Systems.

Hartley, D. S. (1997). Verification & validation in military simulations. In S. Andradottir, K. J. Healy, D. H. Withers, & B. L. Nelson (Eds.), *Proceedings of the 1997 Winter Simulation Conference* (pp. 925–932). Washington, DC.

Hartley, D. S. (1999). *PACOM instability indicators workshop, Y/DSRD-3134*. Oak Ridge, TN: Lockheed Martin Energy Systems.

Hartley, D. S. (2001a). Battle modeling. In S. I. Gass & C. M. Harris (Eds.), *Encyclopedia of operations research and management science, centennial edition* (pp. 53–57). Boston: Kluwer Academic Publishers.

Hartley, D. S. (2001b). Military operations other than war. In S. I. Gass & C. M. Harris (Eds.), *Encyclopedia of operations research and management science, centennial edition* (pp. 508–513). Boston: Kluwer Academic Publishers.

Hartley, D. S. (2001c). *OOTW toolbox specification, DSH-0324*. Oak Ridge, TN: Hartley Consulting.

Hartley, D. S. (2001d). *Predicting combat effects*. Linthicum, MD: INFORMS.

Hartley, D. S. (2003). *Flexible Asymmetric Simulation Technologies (FAST): Beta-1 build test report*. Andover, MA: DRC.

Hartley, D. S. (2004). *FAST for the warfighter: Test strategy & plan (Revision 3)*. Vienna, VA: DRC.

Hartley, D. S. (2005a). *Military Operations Other Than War (MOOTW): Flexible Asymmetric Simulation Technologies (FAST) prototype toolbox: FY05 validation strategy & plan*. Vienna, VA: Dynamics Research Corp.

Hartley, D. S. (2005b). *MOOTW FAST prototype toolbox: FY04 validation strategy & plan*. Vienna, VA: DRC.

Hartley, D. S. (2006a). *Interim semi-static stability model*. Retrieved May 4, 2016, from Hartley Consulting: http://drdeanhartley.com/HartleyConsulting/TOOLBOX/issm.htm

Hartley, D. S. (2006b). *Operations Other Than War (MOOTW) Flexible Asymmetric Simulation Technologies (FAST) prototype toolbox: ISSM v4.00 users' guide*. Orlando, FL: Dynamics Research Corporations.

Hartley, D. S. (2006c). *Operations Other Than War (OOTW) Flexible Asymmetric Simulation Technologies (FAST) prototype toolbox: ISSM v4.00 analysts' guide*. Orlando, FL: Dynamics Research Corporation.

Hartley, D. S. (2008). *DIME/PMESII VV&A tool*. Retrieved May 4, 2016, from Hartley Consulting: http://drdeanhartley.com/HartleyConsulting/VVATool/VVA.htm

Hartley, D. S. (2009). *I2WD HSCB testbed: 2009 V&V strategy & plan*. Ft Monmouth, NJ: US Army RDECOM CERDEC I2WD.

Hartley, D. S. (2010a). Comparing validation results for two DIME/PMESII models: Understanding coverage profiles. In B. Johansson, S. Jain, J. Montoya-Torres, & E. Yucesan (Eds.), *Proceedings of the 2010 Winter Simulation Conference*. IEEE.

Hartley, D. S. (2010b). Corruption in Afghanistan: Conceptual model. In *NDU Corruption Workshop*. Washington, DC: National Defense University.

Hartley, D. S. (2010c). VV&A for DIME/PMESII models. In A. Woodcock, M. Baranick, & A. Sciarretta (Eds.), *The human social culture behavior modeling workshop* (pp. 219–242). Washington, DC: National Defense University.

Hartley, D. S. (2012). *Validation process for the DEXES II conceptual model*. Oak Ridge, TN: Hartley Consulting.

Hartley, D. S. (2015a). DIME/PMESII models. In P. V. Fellman, Y. Bar-Yam, & A. A. Minai (Eds.), *Conflict and complexity: Countering terrorism, insurgency, ethnic and regional violence* (pp. 111–136). New York: Springer.

Hartley, D. S. (2015b, March). Practical VV&A for DIME/PMESII models. *Phalanx, 48*(1).

Hartley, D. S. (2016a). *IW ontology GUI development*. Retrieved May 4, 2016, from Hartley Consulting: http://drdeanhartley.com/HartleyConsulting/IWMetricOntology/IWOntologyGUI.htm

Hartley, D. S. (2016b). *Ontologies*. Retrieved May 19, 2016, from Hartley Consulting: http://drdeanhartley.com/HartleyConsulting/Ontologies.htm

Hartley, D. S., Bell, R., & Packard, S. (1999). *OOTW force design tools, Y/DSRD-3117*. Oak Ridge, TN: Lockheed Martin Energy Systems.

Hartley, D. S., Holdsworth, D., & Farrell, C. (2006). *OOTW FAST prototype toolbox: Analysis process*. Orlando, FL: DRC.

Hartley, D. S., Kruse, K., Denson, J., Austin, G., & Breuer, R. (1989). *Sensitivity analysis of the joint theater level simulation I, K/DSRD-70*. Oak Ridge, TN: Martin Marietta Energy Systems.

Hartley, D. S., Kruse, K. L., Martellaro, A. J., Packard, S. L., Thomas, B. J., & Turley, V. K. (1994). *An independent verification and validation of the future theater level model conceptual model, K/DSRD-1637*. Oak Ridge, TN: Martin Marietta Energy Systems.

Hartley, D. S., & Lacy, L. W. (2011). *Irregular Warfare (IW) metrics ontology final report, TRAC-H-TR-13-020*. Ft Leavenworth, KS: US Army TRAC.

Hartley, D. S., & Lacy, L. W. (2013a). Creating the foundations for modeling irregular warfare. In D. M. Nicholson & D. D. Schmorrow (Eds.), *Advances in design for cross-cultural activities, Part II* (pp. 13–23). Boca Raton, FL: CRC Press.

Hartley, D. S., & Lacy, L. W. (2013b). *IW ontology final report*. Ft Leavenworth, KS: US Army TRAC.

Hartley, D. S., Loebl, A., Rigdon, B., van Leeuwen, B. P., & Harrigan, R. (2004). *Costs of lack of commonality: Initial findings from the commonality pathfinder project*. Oak Ridge, TN: Oak Ridge National Laboratory.

Hartley, D. S., & Packard, S. L. (1998a). *OOTW cost tools, Y/DSRD-3099*. Oak Ridge, TN: Lockheed Martin Energy Systems.

Hartley, D. S., & Packard, S. L. (1998b). *OOTW tool requirements in relation to JWARS, Y/DSRD-3076*. Oak Ridge, TN: Lockheed Martin Energy Systems.

Hartley, D. S., Quillinan, J. D., & Kruse, K. L. (1990). *Phase I verification and validation of SIMNET-T, K/DSRD-116*. Oak Ridge, TN: Martin Marietta Energy Systems.

Hartley, D. S., & Starr, S. (2010). Verification and validation. In A. Kott & G. Citrenbaum (Eds.), *Estimating impact* (pp. 311–336). New York: Springer.

Hartley, D. S., & Whitley, H. (1996). Verification, validation, and certification (VV&C). In C. E. Gettig & J. I. Palmore (Eds.), *Simulation data and its management mini-symposium (SIMDATAM '95)*. Alexandria, VA: MORS.

Haskins, C. (2010, September–October). A practical approach to cultural insight. *Military Review*.

Haut, D. G. (1994). OOTW (D. S. Hartley, Interviewer)

Haut, D. G. (1995). Progress in computer-assisted training simulations for lesser regional contingencies. *JUORS VII, 28 November–1 December 1995*. Tokyo, Japan.

Hayes, B. C., & Sands, J. I. (1997). *Doing windows: Non-traditional military responses to complex emergencies*. Washington, DC: CCRP.

Hayes, M. D., & Wheatley, G. F. (1996). *Interagency and political-military dimensions of peace operations: Haiti—A case study*. Washington, DC: National Defense University.

Hillson, R., et al. (2009). *Requirements for a Government owned DIME/PMESII model suite*. Washington, DC: Office of the Secretary of Defense Modeling & Simulation Steering Committee.

Hoffman, B. (1994). *Responding to terrorism across the technological spectrum, AD-A283 936*. Ft Leavenworth, KS: US Army War College.

Hogg, D. R. (1993). *A military campaign against gangs: Internal security operations in the United States by active duty forces, AD-A274 041*. Ft Leavenworth, KS: US Army CGSC.

HQ DA. (1969). *Special forces operations, FM 31-21*. Washington, DC.

HQ DA. (2006a). *Counterinsurgency, FM 3-24*. Washington, DC.

HQ DA. (2006b). *The operations process, FM 5-0.1*. HQDA.

HQ DA. (2008a, February). *Operations, FM 3-0*. Washington, DC: Headquarters Department of the Army.

HQ DA. (2008b, October). *Stability operations, FM 3-07*. Washington, DC: Headquarters Department of the Army.

Institute for Defense Analyses. (1998). *Contingency operations support tool web site*. Retrieved 1998, from http://www.ida.org/COST

Jackson, J. (2010). Population and social dynamics modeling. In A. Woodcock, M. Baranick, & A. Sciarretta (Eds.), *The human social culture behavior modeling workshop* (p. 279). Washington, DC: National Defense University.

Johnson, N. F. (2013). From particles to people: Social, cultural and behavioral dynamics in insurgency, terrorism and beyond. In D. M. Nicholson & D. D. Schmorrow (Eds.), *Advances in design for cross-cultural activities, Part II* (pp. 398–407). Boca Raton, FL: CRC Press.

Johnson, N. F., Zheng, M., Vorobyeva, Y., Gabriel, A., Qi, H., Velasquez, N., et al. (2016). New online ecology of adversarial aggregates: ISIS and beyone. *Science, 352*(6292), 1459–1463.

Jonker, D., & Wright, W. (2010). Visualization and comprehension. In A. Kott & G. Citrenbaum (Eds.), *Estimating impact* (pp. 285–309). New York: Springer.

King, M. L., Hering, A. S., & Aguilar, O. M. (2016, October). Building predictive models of counterinsurgent deaths using robust clustering and regression. *The Journal of Defense Modeling and Simulation: Applications, Methodology, Technology, 13*(4), 449–465.

King, M. L., Hering, A. S., & Newman, A. M. (2014). Evaluating counterinsurgency classification schemes. *Military Operations Research, 19*(3), 5–25.

Klein, G. (2011). A taxonomy for HSCB research and operations. *HSCB Focus 2011 Conference, February 8–10, 2011*. Chantilly, BA.

Knepell, P. L., & Arangno, D. C. (1993). *Simulation validation: A confidence assessment methodology*. Los Alamitos, CA: IEEE Computer Society Press.

Kott, A., & Citrenbaum, G. (Eds.). (2010). *Estimating impact*. New York: Springer.

Kott, A., & Skarin, B. (2010). Insurgency and security. In A. Kott & G. Citrenbaum (Eds.), *Estimating impact* (pp. 239–262). New York: Springer.

Kuznar, L. A. (2010). Bridging the micro–macro Gulf in social science research: Military and intelligence applications. In A. Woodcock, M. Baranick, & A. Sciarretta (Eds.), *The human social culture behavior modeling workshop* (pp. 137–147). Washington, DC: National Defense University.

Lacy, L. W. (2005). *OWL: Representing information using the web ontology language*. Victoria, BC: Trafford.

Lidy, A. M. (1998). *United States military role in smaller scale contingencies, D2166*. Alexandria, VA: Institute for Defense Analyses.

Lobban, R. A. (2010). Marrying modeling with empiricism: The case and context of Sudan. In A. Woodcock, M. Baranick, & A. Sciarretta (Eds.), *The human social culture behavior modeling workshop* (pp. 184–214). Washington, DC: National Defense University.

Lofdahl, C. (2010). Governance and society. In A. Kott & G. Citrenbaum (Eds.), *Estimating impact* (pp. 179–204). New York: Springer.

Lowrance, J. D., & Murdock, J. L. (2009). *Political, Military, Ecomomic, Social, Infrastructure, Information (PMESII) effects forecasting for Course of Action (COA) evaluation*. Rome, NY: Air Force Research Laboratory.

Marlin, B. (2009). *Ascertaining validity in the abstract realm of PMESII simulation models: An analysis of the Peace Support Operations Model (PSOM)*. Masters Thesis, Naval Postgraduate School, Monterey, CA.

McCurdy, M. (2016, October 27). Theater engagement analysis (D. S. Hartley, Interviewer).

McGrady, K. A., et al. (1995). *CNA's humanitarian assistance operations game: A summary report, CIM 392*. Alexandria, VA: Center for Naval Analyses.

McKenna, P. (2010). The deterrence analytic challenge. In A. Woodcock, M. Baranick, & A. Sciarretta (Eds.), *The human social culture behavior modeling workshop* (pp. 287–297). Washington, DC: National Defense University.

Metz, S. (1993). *The future of insurgency, AD-A274 395*. Carlisle Barracks, PA: US Army War College.

Metz, S. (1994). *Disaster and intervention in sub-Saharan Africa: Learning from Rwanda, AD-A285 320*. Carlisle Barracks, PA: US Army War College.

Moore, W. H. (2010). Cross-national correlates of terror: Empirical analyses of the late 20th century. In A. Woodcock, M. Baranick, & A. Sciarretta (Eds.), *The human social culture behavior modeling workshop* (pp. 65–82). Washington, DC: National Defense University.

National Simulation Center. (1996). *What simulations are available now?* Retrieved April 27, 2016, from http://www.ibrarian.net/navon/paper/PART_III_WHAT_SIMULATIONS_ARE_AVAILABLE_NOW__Chap.pdf?paperid=9972164

Newett, S. (1996). *Planning for humanitarian assistance operations, CRM 95-151*. Alexandria, VA: Center for Naval Analyses.

Newett, S., et al. (1996). *Emerald express '95: Analysis report, CRM 95-156*. Alexandria, VA: Center for Naval Analyses.

NPS SEED. (n.d.). *SEED center for data farming*. Retrieved May 14, 2016, from SEED Center for Data Farming: https://harvest.nps.edu/

Ontology (information science). (n.d.). Retrieved July 6, 2016, from Wikipedia: https://en.wikipedia.org/wiki/Ontology_(information_science)

Ottenberg, M. (2010). Wargaming/gaming. In A. Woodcock, M. Baranick, & A. Sciarretta (Eds.), *The human social culture behavior modeling workshop* (p. 278). Washington, DC: National Defense University.

Partnow, P. H., & Hartley, D. S. (2010). Using cultural information to model DIME/PMESII effects. In A. Woodcock, M. Baranick, & A. Sciarretta (Eds.), *The human social culture behavior modeling workshop* (pp. 18–30). Washington, DC: National Defense University.

Persons, J. (2010). Discussion of HSCB and OR/M&S/Analysis: USA/USMC background study. In A. Woodcock, M. Baranick, & A. Sciarretta (Eds.), *The human social culture behavior modeling workshop* (pp. 270 & 283–284). Washington, DC: National Defense University.

Pfeffer, J., & Carley, K. M. (2013). Social networks, social media, social change. In D. M. Nicholson & D. D. Schmorrow (Eds.), *Advances in design for cross-cultural activities, Part II* (pp. 273–282). Boca Raton, FL: CRC Press.

Pierson, B., Barge, W., & Crane, C. (2010). The hairball that stabilized Iraq: Modeling FM 3-24. In A. Woodcock, M. Baranick, & A. Sciarretta (Eds.), *The human social culture behavior modeling workshop* (pp. 362–372). Washington, DC: National Defense University.

Post, J. M. (2010). When hatred is bred in the bone: Psycho-cultural foundations of contemporary terrorism. In A. Woodcock, M. Baranick, & A. Sciarretta (Eds.), *The human social culture behavior modeling workshop* (pp. 147–163). Washington, DC: National Defense University.

Protégé. (n.d.). Retrieved July 16, 2016, from http://protege.stanford.edu/

Pynadath, D. V., & Marsella, S. C. (2013). Socio-cultural modeling through decision-theoretic agents with theory of mind. In D. M. Nicholson & D. D. Schmorrow (Eds.), *Advances in design for cross-cultural activities, Part II* (pp. 417–426). Boca Raton, FL: CRC Press.

Roland, R. J. (2016, November 6). JTLS (D. S. Hartley, Interviewer)

Rolands & Associates Corporation. (2016). *JTLS-GO menu*. Retrieved November 6, 2016, from Rolands & Associates Corporation: https://www.rolands.com/jtls/j_over.php

Rosbolt, P. K. (1995). *Islamic fundamentalism: Considerations for the operational commander, AD-A298 118*. Newport, RI: Naval War College.

Rose, G., & Lambert, N. (1996). Operational analysis in the front line (Five Go Camping in Bosnia!). *The Thirteenth International Symposium on Military Operational Research (13 ISMOR)*. Shrivenham, UK: The Royal Military College of Science.

Senko, R. M. (2005). *Flexible Asymmetric Simulation Technologies (FAST) prototype toolbox: FY05 final verification test report*. Vienna, VA: Dynamics Research Corp.

Shaw, E. J. (1995). *The U.S. Southern command and the Andean drug war, AD-A293 436*. Newport, RI: Naval War College.

Shedlowski, D. J. (1995). Peacekeeping Operations (PEKO). In *The Japan-US Operations Research Seminar VII (JUORS VII)* (pp. 60–91). Tokyo, Japan: JUORS.

Siegel, A. B. (1995). *Requirements for humanitarian assistance and peace operations: Insights from seven case studies, CRM 94-74*. Alexandria, VA: Center for Naval Analyses.

Siegel, D. A. (2010). Social network structure and counterinsurgency/counterterrorism: Using theory to limit causal links and aid in strategic planning. In A. Woodcock, M. Baranick, & A. Sciarretta (Eds.), *The human social culture behavior modeling workshop* (pp. 83–106). Washington, DC: National Defense University.

Silverman, B. G. (2010). Is systems engineering the new social science? Or, a multi-resolution architecture for analytic studies of the socio-cultural and cognitive layers of a society. In A. Woodcock, M. Baranick, & A. Sciarretta (Eds.), *The human social culture behavior modeling workshop* (pp. 107–116). Washington, DC: National Defense University.

Silverman, B. G. (2012). *StateSim: Socio-cognitive agents in meta-networks*. Retrieved May 14, 2016, from Barry G. Silverman, Ph.D.: http://www.seas.upenn.edu/~barryg/documents/StateSimFactsheet6.pdf

SimCity 2000. (1994). Retrieved May 30, 2016, from Wikipedia: https://en.wikipedia.org/wiki/SimCity_2000

Sokolowski, J. (2010). Counter-insurgency modeling. In A. Woodcock, M. Baranick, & A. Sciarretta (Eds.), *The human social culture behavior modeling workshop* (p. 278). Washington, DC: National Defense University.

Sovereign, M. (1998). *Humanitarian assistance and disaster relief in the next century: Workshop report*. Arlington, VA: National Defense University.

Stahl, D. T. (1992). *Noncombatant evacuation operations in support of the national military strategy, ADA-255 136*. Ft Leavenworth, KS: School of Advanced Military Studies.

Staniec, C., & Hartley, D. S. (1999). *OOTW Analysis and Modeling Techniques (OOTWAMT) Workshop Proceedings, January 1997*. Alexandria, VA: MORS.

Starr, S., Hartley, D. S., Clemence, R. D., McEver, J. D., Noble, D. F., Signori, D. T., et al. (2007). *Verification, validation, and accreditation (VV&A)*. Vienna, VA: Evidence Based Research.

Sun-Tzu. (1963). *The art of war*. (S. B. Griffith, Trans.). New York: Oxford University Press.

The Dupuy Institute. (1995). *Peacekeeping in Bosnia: Fatality estimates*. McLean, VA: The Dupuy Institute.

TRAC. (2009, January). *IW decomposition analytic strategy: Overview briefing for IW WG*. Ft Leavenworth, KS: TRAC.

TRAC. (2010). *Metrics v3.xls*. Ft Leavenworth, KS: TRAC.

Tufte, E. R. (2001). *The visual display of quantitative information* (2nd ed.). Cheshire, CT: Graphics Press.

Unal, M. C. (2016). Counterinsurgency and military strategy: An analysis of the Turkish army's COIN strategies/doctrines. *Military Operations Research, 21*(1), 55–88.

Van Brackle, D., & Haglich, P. (2013). Improvements in the Jabari event coder. In D. M. Nicholson & D. D. Schmorrow (Eds.), *Advances in design for cross-cultural activities, Part II* (pp. 191–200). Boca Raton, FL: CRC Press.

Walker, P. B., Anglero, T., & Davidson, I. (2013). Network discovery: Measuring cause and effect behind event and social networks. In D. M. Nicholson & D. D. Schmorrow (Eds.), *Advances in design for cross-cultural activities, Part II* (pp. 325–334). Boca Raton, FL: CRC Press.

Waltz, E., & Lawrence, C. (2009). Conflict Modeling Planning and Outcomes Experimentation (COMPOEX). *Human social culture behavior modeling workshop II, June 8–10 2009*. Washington, DC: NDU.

Wedgwood, J. E., Ruvinsky, A., & Siedlecki, T. (2013). What lies beneath: Forecast transparency to foster understanding and trust in forecast models. In D. M. Nicholson & D. D. Schmorrow (Eds.), *Advances in design for cross-cultural activities, Part II* (pp. 64–73). Boca Raton, FL: CRC Press.

Wentz, L. (Ed.). (1997). *Lessons from Bosnia: The IFOR experience*. Vienna, VA: CCRP.

Wikipedia. (n.d.). *Influence diagram*. Retrieved May 6, 2016, from Wikipedia: https://en.wikipedia.org/wiki/Influence_diagram

Woodcock, A. E. (1996). Modeling and analysis of societal dynamics: The Deployable Exercise Support (DEXES) System. In A. Woodcock & D. Davis (Eds.), *Analytical approaches to the study of future conflict* (pp. 255–279). Clementsport, Canada: The Canadian Peacekeeping Press.

Woodcock, A., Baranick, M., & Sciarretta, A. (Eds.). (2010). *The human social culture behavior modeling workshop*. Washington, DC: National Defense University.

Yohai, I., McCormack, R., Sidman, J., Galantowicz, J., Borden, C., & Pritchett, S. (2013). Combining social and environmental models. In D. M. Nicholson & D. D. Schmorrow (Eds.), *Advances in design for cross-cultural activities, Part II* (pp. 74–83). Boca Raton, FL: CRC Press.

Yohai, I., Riordan, B., Duchon, A., & Paley, M. (2013). Discovering entity characteristics and relationships through topic modeling. In D. M. Nicholson & D. D. Schmorrow (Eds.), *Advances in design for cross-cultural activities, Part II* (pp. 44–53). Boca Raton, FL: CRC Press.

Youngblood, S. (2005). *VV&A standards initiative*. Washington, DC: DMSO.

Youngblood, S. M., Pace, D. K., Eirich, P. L., Gregg, D. M., & Coolahan, J. E. (2000). *Simulation verification, validation, and accreditation*. Retrieved May 22, 2016, from John Hopkins APL Technical Digest: http://www.jhuapl.edu/techdigest/TD/td2103/youngblood.pdf

Zvijac, D. J., & McGrady, K. A. (1994). *Operation restore hope: Summary report*. Alexandria, VA: Center for Naval Analyses.

Index

D.S. Hartley III, *Unconventional Conflict*, Understanding Complex Systems,
DOI 10.1007/978-3-319-51935-7

FSC
www.fsc.org
MIX
Papier | Fördert gute Waldnutzung
FSC® C083411

Kolibri 360 GmbH
Ferdinand-Jühlke-Straße 7
99095 Erfurt, Deutschland
produktsicherheit@kolibri360.de